Partial Differential Equations

Mathematical Engineering, Manufacturing, and Management Sciences

Series Editor
Mangey Ram
Professor, Assistant Dean (International Affairs), Department of Mathematics
Graphic Era University, Dehradun, India

The aim of this new book series is to publish the research studies and articles that bring up the latest development and research applied to mathematics and its applications in the manufacturing and management sciences areas. Mathematical tools and techniques are the strength of engineering sciences. They form the common foundation of all novel disciplines as engineering evolves and develops. The series will include a comprehensive range of applied mathematics and its application in engineering areas such as optimization techniques, mathematical modelling and simulation, stochastic processes and systems engineering, safety-critical system performance, system safety, system security, high assurance software architecture and design, mathematical modelling in environmental safety sciences, finite element methods, differential equations, reliability engineering, and so on.

Recent Advancements in Graph Theory
Edited by N. P. Shrimali and Nita H. Shah

Mathematical Modeling and Computation of Real-Time Problems:
An Interdisciplinary Approach
Edited by Rakhee Kulshrestha, Chandra Shekhar, Madhu Jain, and Srinivas R. Chakravarthy

Circular Economy for the Management of Operations
Edited by Anil Kumar, Jose Arturo Garza-Reyes, and Syed Abdul Rehman Khan

Partial Differential Equations: An Introduction
Nita H. Shah and Mrudul Y. Jani

Linear Transformation: Examples and Solutions
Nita H. Shah and Urmila B. Chaudhari

Matrix and Determinant: Fundamentals and Applications
Nita H. Shah and Foram A. Thakkar

Non-Linear Programming: A Basic Introduction
Nita H. Shah and Poonam Prakash Mishra

For more information about this series, please visit: https://www.routledge.com/ Mathematical-Engineering-Manufacturing-and-Management-Sciences/book-series/ CRCMEMMS

Partial Differential Equations
An Introduction

Nita H. Shah and Mrudul Y. Jani

CRC Press
Taylor & Francis Group
Boca Raton London New York

CRC Press is an imprint of the
Taylor & Francis Group, an **informa** business

First edition published 2021
by CRC Press
6000 Broken Sound Parkway NW, Suite 300, Boca Raton, FL 33487-2742

and by CRC Press
2 Park Square, Milton Park, Abingdon, Oxon, OX14 4RN

ISBN: 9780367613228 (hbk)
ISBN: 9781003105183 (ebk)

Typeset in Times
by MPS Limited, Dehradun

Contents

Acknowledgements

First, at this stage, I would like to extend my sincere gratitude and thank my PhD guide Prof. (Dr.) Nita H. Shah for her constant encouragement and support which I cannot describe in words. She has been an inspiration for me. I have witnessed her great multidisciplinary knowledge and enthusiasm. From her, I have learned to be dedicated, energetic, punctual, sharp, and patient.

I express heartfelt gratitude to my loving wife Dr. Urmila for her positive suggestions, and continuous motivation to improve the standard of this book. Her unconditional support has made my journey of writing this book a satisfactory success, which I will cherish forever in my life. I am also very thankful to my mother Purnaben and father Yogeshkumar for their constant support.

Dr. Mrudul Y. Jani

Preface

Differential equations play a noticeable role in physics, engineering, economics, and other disciplines. Differential equations permit us to model changing forms in both mathematical and physical problems. These equations are precisely used when a deterministic relation containing some continuously varying quantities and their rates of change in space and/or time is recognized or postulated. So, the study of partial differential equations is much more important in comparison with ordinary differential equations. The important role of partial differential equations is precisely replicated by the fact that it is the motivation of all the books on the differential equation. The partial differential equation can be used to describe an extensive variety of occurrences, such as heat, fluid dynamics, diffusion, quantum mechanics, sound, elasticity, electrostatics, and electrodynamics. Therefore, we turn our attention to the area of partial differential equations.

Authors

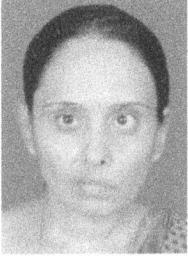

Nita H. Shah, PhD, received her PhD in statistics from Gujarat University in 1994. From February 1990 until now, Prof. Shah has been HOD of the Department of Mathematics at Gujarat University, India. She is a post-doctoral visiting research fellow at the University of New Brunswick, Canada. Prof. Shah's research interests include inventory modelling in supply chains, robotic modelling, mathematical modelling of infectious diseases, image processing, dynamical systems and their applications, and so on. She has published 13 monographs, 5 textbooks and 475+ peer-reviewed research papers. She has edited four books with coeditor Dr. Mandeep Mittal, which were published by IGI-Global and Springer; and papers, which have appeared in Elsevier, Inderscience, and Taylor & Francis journals. She is the author of 14 books. The total number of citations is over 3064 and the maximum number of citations for a single paper is over 174 on Google Scholar. Prof. Shah's H-index is 24 and the her i-10 index is 77. She has guided 28 PhD students and 15 MPhil students. Seven students are pursuing research for their PhD degrees. She has travelled in the United States, Singapore, Canada, South Africa, Malaysia, and Indonesia to give talks. She is the vice-president of the Operational Research Society of India and is a council member of the Indian Mathematical Society.

Mrudul Y. Jani, PhD, is an associate professor in the Department of Applied Sciences and Humanities, PIET, Faculty of Engineering and Technology at Parul University, Vadodara, Gujarat, India. He has 11+ years of teaching experience and more than 5 years of research experience in the field of inventory management. His research interests include inventory management under deterioration and different demand structures. Dr. Jani has had over 25+ articles and book chapters published in international journals, published by TOP-Springer, Revista Investigacion Operacional, JOBARI, IJIEC, Taylor & Francis, Springer, IGI Global, Growing Sciences, IJOQM, AMSE, IJSCM, IJBPSCM, OSCM, IJLSM, TWMS, and IJMOR. He has also written one book for international publication.

1 Introduction of Partial Differential Equations

Partial differential equations arise quite often in numerous engineering and physical problems when the functions contain two or more independent variables. Several problems in fluid mechanics and solid mechanics, electromagnetic theory, heat transfer, vibrations, and many other thrust areas of engineering lead to the study of partial differential equations.

While solving the problems of an ordinary differential equation, one should find the general solution first and then determine the arbitrary constants using the initial conditions and finally evaluate the particular solution. But, the same method cannot be applicable in the case of partial differential equations. Instead, in most of the problems of partial differential equations in a region, initial conditions are used to get the particular solution; and boundary conditions are used to evaluate the arbitrary constants or arbitrary functions at the boundary of the region.

In this chapter, we will study the definition of partial differential equations with some examples, Order of partial differential equations, the formation of partial differential equations, and the Direct Integration Method to solve some particular types of partial differential equations.

1.1 PARTIAL DIFFERENTIAL EQUATIONS

A differential equation that contains two or more independent variables is called a **partial differential equation**. A partial differential equation for the function $z(x_1, \dots x_n)$ is an equation of the form $f\left(x_1, \dots x_n; \frac{\partial z}{\partial x_1}, \dots \frac{\partial z}{\partial x_n}; \frac{\partial^2 z}{\partial x_1 \partial x_1}, \dots \frac{\partial^2 z}{\partial x_1 \partial x_n}; \dots\right) = 0.$

Some standard types of partial differential equations are

$\frac{\partial u}{\partial t} = a^2 \frac{\partial^2 u}{\partial x^2}$ One-dimensional heat equation

$\frac{\partial^2 u}{\partial t^2} = a^2 \frac{\partial^2 u}{\partial x^2}$ One-dimensional wave equation

$u_{xx} + u_{yy} = 0$ Two-dimensional Laplace equation

$u_{xx} + u_{yy} + u_{zz} = 0$ Three-dimensional Laplace equation

The **order** of a partial differential equation is the order of the highest derivatives in the equation. The order of the above equations is 2.

Whereas, the order of $\frac{\partial z}{\partial x} + 2\frac{\partial z}{\partial y} = 5$ is one.

Usual Notations: If $z = f(x, y)$ be a function of two independent variables x and y then we use following usual notations for partial derivatives,

$$p = \frac{\partial z}{\partial x}, \quad q = \frac{\partial z}{\partial y}, \quad r = \frac{\partial^2 z}{\partial x^2}, \quad s = \frac{\partial^2 z}{\partial x \partial y}, \quad t = \frac{\partial^2 z}{\partial y^2}.$$

Verification of Solution of Partial Differential Equations

Example 1.1: Verify that the equation $u = e^x \cos y$ is the solution to Laplace's equation $\frac{\partial^2 u}{\partial x^2} + \frac{\partial^2 u}{\partial y^2} = 0$ or not.

Solution: Given equation is $u = e^x \cos y$.

Now, let $\frac{\partial u}{\partial x} = e^x \cos y \Rightarrow \frac{\partial^2 u}{\partial x^2} = e^x \cos y$ and

let $\frac{\partial u}{\partial y} = -e^x \sin y \Rightarrow \frac{\partial^2 u}{\partial y^2} = -e^x \cos y$

$$
\begin{aligned}
\text{L. H. S.} \quad &= \frac{\partial^2 u}{\partial x^2} + \frac{\partial^2 u}{\partial y^2} \\
&= e^x \cos y - e^x \cos y \\
&= 0 \\
&= \text{R. H. S.}
\end{aligned}
$$

Hence the proof.

1.2 FORMATION OF PARTIAL DIFFERENTIAL EQUATIONS

The partial differential equations can be formed in two different ways:

I.   Elimination of arbitrary constants that are present in the functional relationship between variables.
II. Elimination of arbitrary functions from the given relations.

I.   By Eliminating Arbitrary Constants

Note: The number of arbitrary constants in the functional relation is equal to the number of times partial derivative one has to take to obtain the partial differential equation.

Consider, the function $f(x, y, z, a, b) = 0$. Where, a and b are independent arbitrary constants.

Step 1:

$$f(x, y, z, a, b) = 0 \qquad\qquad (1.1)$$

Step 2:

$$\text{Find } \frac{\partial f}{\partial x} = 0 \text{ and } \frac{\partial f}{\partial y} = 0 \qquad\qquad (1.2)$$

Step 3: Eliminating a and b from equations (1.1) and (1.2), the partial differential equation of form $F(x, y, z, p, q) = 0$ can be obtained.

II.   By Eliminating Arbitrary Functions

Note: The number of arbitrary functions is equal to the order of partial differential equations.

Consider, the function either in the form $f(u, v) = 0$ or $v = f(u)$.

Step 1: Find $\frac{\partial u}{\partial x}, \frac{\partial u}{\partial y}, \frac{\partial v}{\partial x}$ and $\frac{\partial v}{\partial y}$

Step 2: Find Jacobian $J = \dfrac{\partial(u, v)}{\partial(x, y)} = \begin{vmatrix} \dfrac{\partial u}{\partial x} & \dfrac{\partial u}{\partial y} \\ \dfrac{\partial v}{\partial x} & \dfrac{\partial v}{\partial y} \end{vmatrix}$

Step 3: Equate $\begin{vmatrix} \dfrac{\partial u}{\partial x} & \dfrac{\partial u}{\partial y} \\ \dfrac{\partial v}{\partial x} & \dfrac{\partial v}{\partial y} \end{vmatrix} = 0$ and the partial differential equation of the form

$F(x, y, z, p, q) = 0$ can be obtained.

Example 1.2: Eliminate the constants a and b from $z = (x + a)(y + b)$.

Solution: Taking partial derivatives w.r.t., x and y,

$$p = y + b, \quad q = x + a.$$

Substitute $p = y + b$ and $q = x + a$ in $z = (x + a)(y + b)$ and eliminate a and b; the partial differential equation is $z = pq$.

Example 1.3: Form a partial differential equation by eliminating a, b, and c from $\frac{x^2}{a^2} + \frac{y^2}{b^2} + \frac{z^2}{c^2} = 1$.

Solution: Given $\frac{x^2}{a^2} + \frac{y^2}{b^2} + \frac{z^2}{c^2} = 1$

Taking partial derivatives w.r.t., x and y,

$$\frac{2x}{a^2} + \frac{2z}{c^2}p = 0,$$

$$\frac{2y}{b^2} + \frac{2z}{c^2}q = 0 \tag{1.3}$$

Taking partial derivative of equation (1.3) w.r.t., y,

$$0 + \frac{2}{c^2}(zs + qp) = 0 \Rightarrow zs + qp = 0.$$

NOTE: In this problem, more than one partial differential equations are possible. These partial differential equations are $yzt + yq^2 - zq = 0$, $xzr + xp^2 - zp = 0$.

Example 1.4: Eliminate function f from the relation $f(xy + z^2, x + y + z) = 0$.

Solution: Let $u = xy + z^2$ and $v = x + y + z$

Taking partial derivatives of u and v w.r.t., x and y,

$$u_x = y + 2zp, \quad u_y = x + 2zq, \quad v_x = 1 + p, \quad v_y = 1 + q$$

Let, $J = \dfrac{\partial(u, v)}{\partial(x, y)} = \begin{vmatrix} u_x & u_y \\ v_x & v_y \end{vmatrix} = 0 \Rightarrow \begin{vmatrix} y + 2zp & x + 2zq \\ 1 + p & 1 + q \end{vmatrix} = 0$

$$\Rightarrow \frac{1+p}{1+q} = \frac{y+2zp}{x+2zq}$$

Example 1.5: By eliminating an arbitrary function ϕ form the partial differential equation from the relation $xyz = \phi(x + y + z)$.

Solution: Let $u = x + y + z$ and $v = xyz$

Let, $J = \dfrac{\partial(u, v)}{\partial(x, y)} = \begin{vmatrix} u_x & u_y \\ v_x & v_y \end{vmatrix} = 0 \Rightarrow \begin{vmatrix} 1 + p & 1 + q \\ yz + xyp & xz + xyq \end{vmatrix} = 0$

$$\Rightarrow \frac{1+p}{1+q} = \frac{yz+xyp}{xz+xyq}$$

Example 1.6: Eliminate the arbitrary functions f and g from the relation $z = f(x + ay) + g(x - ay)$.

Solution

Hint: In the given problem, a number of functions are two so the order of partial differential equation is two.

Taking partial derivatives of $z = f(x + ay) + g(x - ay)$ w.r.t., x and y,

$$\frac{\partial z}{\partial x} = f'(x + ay) + g'(x - ay); \quad \frac{\partial z}{\partial y} = af'(x + ay) - ag'(x - ay)$$

Once again take partial derivatives w.r.t., x and y,

$$\frac{\partial^2 z}{\partial x^2} = f''(x + ay) + g''(x - ay); \quad \frac{\partial^2 z}{\partial y^2} = a^2 f''(x + ay) + a^2 g''(x - ay)$$

From the above second-order partial derivatives, the resultant partial differential equation is

$$\frac{\partial^2 z}{\partial y^2} = a^2 \frac{\partial^2 z}{\partial x^2}.$$

Example 1.7: Eliminate the arbitrary functions f and ϕ from the relation $z = f(x) + e^y g(x)$.

Solution: Taking partial derivatives of $z = f(x) + e^y g(x)$ w.r.t., x and y,

$$\frac{\partial z}{\partial x} = f'(x) + e^y g'(x); \quad \frac{\partial z}{\partial y} = e^y g(x)$$

Now, substitute $e^y g(x) = \frac{\partial z}{\partial y}$ in $z = f(x) + e^y \phi(x)$

So, $z = f(x) + \frac{\partial z}{\partial y}$

Take once again partial derivative w.r.t., y.

The resultant partial differential equation is $\frac{\partial z}{\partial y} = \frac{\partial^2 z}{\partial y^2}$.

1.3 SOLUTION OF PARTIAL DIFFERENTIAL EQUATIONS

A relation between dependent variables and independent variables which satisfies the partial differential equation is called a **solution of a partial differential equation**. It is also called the **integral of a partial differential equation**.

Complete Solution or Complete Integral

A solution that contains an equal number of arbitrary constants and independent variables is called a complete solution or complete integral.

Particular Solution

In a complete solution by substituting the particular values of the arbitrary constants one can obtain a particular solution.

Singular Solution

If $f(x, y, z, a, b) = 0$ is the complete solution of the partial differential equation $F(x, y, z, p, q) = 0$ then eliminate a and b by taking $\frac{\partial f}{\partial a} = 0$, $\frac{\partial f}{\partial b} = 0$, if it exists, is called a singular solution.

General Solution

In the complete solution $f(x, y, z, a, b) = 0$, take assumption $b = \phi(a)$, so it can be written as

$$f(x, y, z, a, \phi(a)) = 0.$$

Take differentiation of $f(x, y, z, a, b) = 0$ w.r.t., a,

$$\frac{\partial f}{\partial a} + \frac{\partial f}{\partial b}\phi'(a) = 0.$$

Eliminating 'a' from $f(x, y, z, a, \phi(a)) = 0$ and $\frac{\partial f}{\partial a} + \frac{\partial f}{\partial b}\phi'(a) = 0$, if it exists, is called a general solution of $F(x, y, z, p, q) = 0$.

1.3.1 DIRECT INTEGRATION METHOD TO SOLVE PARTIAL DIFFERENTIAL EQUATIONS

The partial differential equations which contain only a single partial derivative term can be solved using this method.

Example 1.8: Solve $\frac{\partial^2 z}{\partial x \partial y} = x^2 + y^2$

Solution: Given a partial differential equation is $\frac{\partial}{\partial x}\left(\frac{\partial z}{\partial y}\right) = x^2 + y^2$.

Take integration on both the sides w.r.t., x and consider y as a constant.

$$\frac{\partial z}{\partial y} = \frac{x^3}{3} + xy^2 + f(y)$$

Now, integrating both the sides w.r.t., y and consider x as a constant.

where $F(y) = \int f(y)\, dy$

Example 1.9: Solve $\frac{\partial^2 z}{\partial x \partial y} = \sin x \sin y$, given that $\frac{\partial z}{\partial y} = -2\sin y$, when $x = 0$ and $z = 0$, when y is an odd multiple of $\frac{\pi}{2}$.

Solution: Given a partial differential equation is $\frac{\partial}{\partial x}\left(\frac{\partial z}{\partial y}\right) = \sin x \sin y$.

Take integration on both the sides w.r.t., x and consider y as a constant.

$$\frac{\partial z}{\partial y} = -\cos x \sin y + f(y)$$

Now given that when $x = 0 \Rightarrow \frac{\partial z}{\partial y} = -2\sin y$

$$\therefore \; -2\sin y \;=\; -\cos 0 \sin y + f(y)$$
$$\Rightarrow \; -2\sin y \;=\; -\sin y + f(y)$$
$$\Rightarrow f(y) \qquad = \; -\sin y$$

$$\therefore \; \frac{\partial z}{\partial y} = -\cos x \sin y - \sin y$$

Now, integrating both the sides w.r.t., y and consider x as a constant.

$$\therefore z = \cos x \cos y + \cos y + g(x)$$

Now, it is given that when y is an odd multiple of $\frac{\pi}{2}$ then $z = 0$.

That means if $y = (2k + 1)\frac{\pi}{2}$, $k = 0, \pm1, \pm2 \dots$ then $z = 0$

$$\therefore \; 0 \quad = \cos x \cos(2k+1)\frac{\pi}{2} + \cos(2k+1)\frac{\pi}{2} + g(x)$$
$$\Rightarrow g(x) \; = \; 0$$

$\therefore z = \cos x \cos y + \cos y$ is the required solution.

Example 1.10: Solve $\frac{\partial^2 z}{\partial x^2} = z$

Solution: Given partial differential equation is $\frac{\partial^2 z}{\partial x^2} - z = 0$.

Now, in the given problem, x is the only independent variable.

So, this partial differential equation can be considered as a second-order homogeneous ordinary differential equation with constant coefficients.

$$\text{i.e.} \quad \frac{d^2 z}{dx^2} - z = 0$$

So, the operator form is $(D^2 - 1)z = 0$, where $D = \frac{d}{dx}$.

Now, the auxiliary equation is $m^2 - 1 = 0$.
$\therefore m = \pm1$ are the roots which are real and distinct.

Therefore, the general solution is $z = f_1(y)e^x + f_2(y)e^{-x}$.

EXERCISES

Q1 Verify that the equation $u = \log(x^2 + y^2)$ is the solution to Laplace's equation

$\frac{\partial^2 u}{\partial x^2} + \frac{\partial^2 u}{\partial y^2} = 0$ or not.

Q2 Verify that the equation $u = \tan^{-1}\left(\frac{y}{x}\right)$ is the solution to Laplace's equation $\frac{\partial^2 u}{\partial x^2} + \frac{\partial^2 u}{\partial y^2} = 0$ or not.

Q3 Verify that the equation $u = \sin 9t \sin\left(\frac{x}{4}\right)$ is the solution of a one-dimensional wave equation $\frac{\partial^2 u}{\partial t^2} = a^2 \frac{\partial^2 u}{\partial x^2}$ for any suitable value of a.

Q4 Verify that the equation $u(x, t) = f(x + at) + g(x - at)$ is the solution of a one-dimensional wave equation $\frac{\partial^2 u}{\partial t^2} = a^2 \frac{\partial^2 u}{\partial x^2}$ or not.

Q5 Verify that the equation $u = e^{-2t}\cos x$ is the solution of the heat equation $\frac{\partial u}{\partial t} = a^2 \frac{\partial^2 u}{\partial x^2}$ for any suitable value of a.

Q6 Form partial differential equation from $z = (x - 2)^2 (y - 3)^2$.

Q7 Form partial differential equation from $z = (x^2 + a)(y^2 + b)$.

Q8 By eliminating an arbitrary function f form the partial differential equation from the relation $z = xy + f(x^2 + y^2)$.

Q9 By eliminating an arbitrary function ϕ form the partial differential equation from the relation $z = f\left(\frac{x}{y}\right)$.

Q10 By eliminating an arbitrary function f form the partial differential equation from the relation $f(x^2 - y^2, xyz) = 0$.

Q11 By eliminating an arbitrary function ϕ form the partial differential equation from the relation $\phi(x + y + z, x^2 + y^2 + z^2) = 0$.

Q12 By eliminating the arbitrary functions f and g form the partial differential equation from the relation $z = xf(x + t) + g(x + t)$.

Q13 Solve $\frac{\partial^2 u}{\partial x \partial y} = e^{-y}\cos x$.

Q14 Solve $\frac{\partial^2 z}{\partial x^2} = -z$, given that when $x = 0$ then $z = e^y$ and $\frac{\partial z}{\partial x} = 1$.

Q15 Solve $\frac{\partial^3 u}{\partial x^2 \partial y} = \cos(2x + 3y)$.

ANSWERS

1 Yes
2 Yes
3 For $a = 36$, u is the solution
4 Yes
5 For $a = \sqrt{2}$, u is the solution
6 $4z = p^2 + q^2$
7 $pq = 4xyz$

8 $\frac{p-y}{q-x} = \frac{x}{y}$

9 $\frac{p}{q} = \frac{-y}{x}$

10 $\frac{yz+xyp}{xz+xyq} = \frac{-x}{y}$

11 $\frac{1+p}{1+q} = \frac{x+zp}{y+zq}$

12 $2\frac{\partial^2 z}{\partial x \partial t} = \frac{\partial^2 z}{\partial x^2} + \frac{\partial^2 z}{\partial t^2}$

13 $u(x, y) = -e^{-y}\sin x + F(y) + g(x)$

14 $z = e^y \cos x + \sin x$

15 $u(x, y) = \frac{-\sin(2x+3y)}{12} + xF(y) + G(y) + h(x)$

2 First-Order Partial Differential Equations

A differential equation involving first-order partial derivatives of the dependent variables with respect to independent variables is called a partial differential equation of the first order. The equation can be written in the form, $\phi(x_1, x_2, \ldots x_n, u_{x_1}, u_{x_2}, \ldots u_{x_n}) = 0$. In this chapter, we shall study two main concepts of partial differential equations namely first-order linear partial differential equations and first-order non-linear partial differential equations. In the section on linear first-order partial differential equations, we shall learn the solution procedure of Lagrange's equation of the first order with a sufficient number of examples. In the section on non-linear first-order partial differential equations, we shall discuss two different types of equations. In the section on non-linear first-order partial differential equations, the Charpit auxiliary equation will be derived with some illustrations. In the section on non-linear first-order partial differential equations, some special cases of first-order non-linear partial differential equations are studied with examples. Lastly, some practice examples are given in the exercises.

2.1 LINEAR FIRST-ORDER PARTIAL DIFFERENTIAL EQUATIONS

If first-order partial derivatives occur in first degree only, the dependent variable is not multiplied with its partial derivatives, and partial derivatives are not multiplied with each other then it is called a linear partial differential equation of the first order.

2.1.1 LAGRANGE'S LINEAR EQUATION OF THE FIRST ORDER

Partial differential equations of the form

$$Pp + Qq = R$$

where P, Q, and R are functions of x, y, z or constants, p denotes $\frac{\partial z}{\partial x}$, and q denotes $\frac{\partial z}{\partial y}$ is called the Lagrange linear equation of the first order.

To solve the Lagrange linear equation $Pp + Qq = R$, we will follow the following algorithm.

Step 1: Form the auxiliary equation $\frac{dx}{P} = \frac{dy}{Q} = \frac{dz}{R}$.

Step 2: The auxiliary equations can be solved using the grouping method or the multiplier method (described below) or both to get two independent solutions of auxiliary equations, denoted by

$u(x, y, z) = c_1$, $v(x, y, z) = c_2$; where c_1 and c_2 are arbitrary constants.

11

Step 3: Then $F(u,v) = 0$ *or* $u = f(v)$ is the general solution of the given equation

$$Pp + Qq = R.$$

- **Grouping Method**
  In this method, we compare any two functions which make integration possible. In other words, to complete the first two fractions, the remaining third variable must be absent from them or it is possible to eliminate it by appropriate operations.
- **Multipliers Method**

  In this method, we find two sets of multiplier l, m, n and l', m', and n' either constant or functions of x, y, z such that

$$lP + mQ + nR = 0 \text{ and } l'P + m'Q + n'R = 0$$

Or the selection makes the integration possible.

Remark: Whenever it is required to apply the multipliers method, always give the priority to x, y, z as multipliers. If they don't work, then search for the others.

Example 2.1: Solve $x(y^2 - z^2)p + y(z^2 - x^2)q = z(x^2 - y^2)$.

Solution: The given equation is of the form $Pp + Qq = R$.

Where, $P = x(y^2 - z^2)$, $\quad Q = y(z^2 - x^2)$, $\quad R = z(x^2 - y^2)$.

Now, the Langrage auxiliary equations are

$$\frac{dx}{P} = \frac{dy}{Q} = \frac{dz}{R}$$

$$\therefore \quad \frac{dx}{x(y^2 - z^2)} = \frac{dy}{y(z^2 - x^2)} = \frac{dz}{z(x^2 - y^2)}$$

Using x,y,z as multipliers, we get each ratio as

$$\frac{dx}{x(y^2 - z^2)} = \frac{dy}{y(z^2 - x^2)} = \frac{dz}{z(x^2 - y^2)}$$

$$= \frac{xdx + ydy + zdz}{x^2y^2 - x^2z^2 + y^2z^2 - x^2y^2 + x^2z^2 - y^2z^2}$$

From the last two fractions, we get

$$\frac{dz}{z(x^2 - y^2)} = \frac{xdx + ydy + zdz}{0}$$

$$\Rightarrow xdx + ydy + zdz = 0$$

Integrating, we get

$$\frac{x^2}{2} + \frac{y^2}{2} + \frac{z^2}{2} = \frac{c_1}{2}$$

$$\Rightarrow x^2 + y^2 + z^2 = c_1$$

Thus, $u(x, y, z) = x^2 + y^2 + z^2 = c_1$

Now, for the second solution, we proceed as follows. Using $\frac{1}{x}, \frac{1}{y}, \frac{1}{z}$ as a multiplier for each of the ratios in the Langrage auxiliary equations, respectively, we get

$$\frac{dx}{x(y^2 - z^2)} = \frac{dy}{y(z^2 - x^2)} = \frac{dz}{z(x^2 - y^2)} = \frac{\frac{dx}{x} + \frac{dy}{y} + \frac{dz}{z}}{y^2 - z^2 + z^2 - x^2 + x^2 - y^2}$$

Comparing the last two fractions, we get

$$\frac{dz}{z(x^2 - y^2)} = \frac{\frac{dx}{x} + \frac{dy}{y} + \frac{dz}{z}}{0}$$

$$\Rightarrow \frac{dx}{x} + \frac{dy}{y} + \frac{dz}{z} = 0$$

Integrating, we get

$$\Rightarrow \log x + \log y + \log z = \log c_2$$

$$\Rightarrow \log xyz = \log c_2$$

$$\Rightarrow xyz = c_2$$

Thus, $v(x, y, z) = xyz = c_2$

Hence, the complete solution is given by

$$F(u, v) = 0 \Rightarrow F(x^2 + y^2 + z^2, xyz) = 0.$$

Example 2.2: Solve $x^2p + y^2q = (x + y)z$

Solution: The given equation is of the form $Pp + Qq = R$.

Where, $P = x^2$, $Q = y^2$, $R = z(x + y)$

Now, the Langrage auxiliary equations are

$$\frac{dx}{P} = \frac{dy}{Q} = \frac{dz}{R}$$

$$\therefore \frac{dx}{x^2} = \frac{dy}{y^2} = \frac{dz}{z(x + y)}$$

From the first two fractions, we get

$$\frac{dx}{x^2} = \frac{dy}{y^2}$$

Integrating, we get

$$-\frac{1}{x} = -\frac{1}{y} + c_1$$

$$\Rightarrow \frac{1}{y} - \frac{1}{x} = c_1$$

$$\Rightarrow \frac{x - y}{xy} = c_1$$

Hence,

$$u(x, y) = \frac{x - y}{xy} = c_1$$

For the second solution using $\frac{1}{x}, \frac{1}{y}, -\frac{1}{z}$ as multipliers, we get

$$\frac{dx}{x^2} = \frac{dy}{y^2} = \frac{dz}{z(x + y)} = \frac{\frac{dx}{x} + \frac{dy}{y} - \frac{dz}{z}}{x + y - (x + y)}$$

Comparing the first and last fractions, we get

$$\frac{dx}{x^2} = \frac{\frac{dx}{x} + \frac{dy}{y} - \frac{dz}{z}}{0}$$

$$\Rightarrow \frac{dx}{x} + \frac{dy}{y} - \frac{dz}{z} = 0$$

Integrating, we get

$$\log x + \log y - \log z = \log c_2$$

$$\Rightarrow \log\left(\frac{xy}{z}\right) = \log c_2$$

$$\Rightarrow \frac{xy}{z} = c_2$$

$$\therefore v(x, y, z) = \frac{xy}{z} = c_2$$

Hence, the complete solution

$$F(u, v) = 0$$

$$\Rightarrow F\left(\frac{x - y}{xy}, \frac{xy}{x}\right) = 0$$

Example 2.3: Solve $pz - qz = (x + y)^2 + z^2$

Solution: The given equation is of the form $Pp + Qq = R$.

Where, $P = z$, $Q = -z$, $R = (x + y)^2 + z^2$

Now, the Langrage auxiliary equations are

$$\frac{dx}{P} = \frac{dy}{Q} = \frac{dz}{R}$$

$$\therefore \frac{dx}{z} = \frac{dy}{-z} = \frac{dz}{(x + y)^2 + z^2}$$

From the first two fractions, we get

$$\therefore \frac{dx}{z} = \frac{dy}{-z}$$

Integrating, we get

$$x + y = c_1$$

Thus, $u(x, y) = x + y = c_1$

Again using the last two fractions

$$\therefore \frac{dy}{-z} = \frac{dz}{(x + y)^2 + z^2}$$

Now, we have $x + y = c_1$

$$\therefore \frac{dy}{-z} = \frac{dz}{z^2 + c_1^2}$$

$$\Rightarrow dy = \frac{-zdz}{z^2 + c_1^2}$$

$$\Rightarrow -2dy = \frac{2zdz}{z^2 + c_1^2}$$

Integrating, we get

$$- 2y = \log(z^2 + c_1{}^2) - c_2$$

$$\Rightarrow \log(z^2 + c_1{}^2) + 2y = c_2$$

$$\Rightarrow \log(z^2 + (x + y)^2) + 2y = c_2$$

Thus, $v(x, y, z) = \log(z^2 + (x + y)^2) + 2y = c_2$

Hence, the complete solution is given by

$$F[x + y, \ \log(z^2 + (x + y)^2) + 2y] = 0$$

Example 2.4: Solve $\frac{y^2z}{x}\frac{\partial z}{\partial x} + xz\frac{\partial z}{\partial y} = y^2$

Solution: Given equation can be expressed as

$$y^2zp + x^2zq = xy^2$$

The given equation is of the form $Pp + Qq = R$.

Where, $P = y^2z$, $Q = x^2z$, and $R = xy^2$

Now, the Langrage auxiliary equations are

$$\frac{dx}{P} = \frac{dy}{Q} = \frac{dz}{R}$$

$$\therefore \ \frac{dx}{y^2z} = \frac{dy}{x^2z} = \frac{dz}{xy^2}$$

From the first two fractions, we get

$$x^2dx = y^2dy$$

Integrating both the sides, we get

$$x^3 - y^3 = c_1$$

Thus, $u(x, y) = x^3 - y^3 = c_1$

Now, considering the first and third term, we have

$$x \, dx = z \, dz.$$

Integrating both the sides, we get

$$x^2 - z^2 = c_2$$

Thus, $v(x, z) = x^2 - z^2 = c_2$

Hence, the complete solution is given by

$$F[x^3 - y^3, x^2 - z^2] = 0$$

Example 2.5: Solve $(y^2 + z^2)\frac{\partial z}{\partial x} - xy\frac{\partial z}{\partial y} + xz = 0$.

Solution: The given equation is of the form $Pp + Qq = R$.

Where, $P = y^2 + z^2$, $Q = -xy$, and $R = -xz$

Now, the Langrage auxiliary equations are

$$\frac{dx}{P} = \frac{dy}{Q} = \frac{dz}{R}$$

$$\therefore \frac{dx}{y^2 + z^2} = \frac{dy}{-xy} = \frac{dz}{-xz}$$

From the last two fractions, we get

$$\therefore \frac{dy}{y} = \frac{dz}{z}$$

Integrating both the sides, we get

$$\log y = \log z + \log c_1$$

$$\therefore \frac{y}{z} = c_1$$

Hence, $u(y, z) = \frac{y}{z} = c_1$

For the second solution using x, y, and z as multipliers, we get

$$\therefore \frac{dx}{y^2 + z^2} = \frac{dy}{-xy} = \frac{dz}{-xz} = \frac{xdx + ydy + zdz}{x(y^2 + z^2) - xy^2 - xz^2}$$

Comparing the last two fractions, we have

$$\therefore \frac{dz}{-xz} = \frac{xdx + ydy + zdz}{0}$$

$$\therefore xdx + ydy + zdz = 0$$

Integrating both the sides, we have

$$\therefore x^2 + y^2 + z^2 = c_2$$

Hence, $v(x, y, z) = x^2 + y^2 + z^2 = c_2$

Hence, the complete solution is given by

$$F\left[\frac{y}{z}, x^2 + y^2 + z^2\right] = 0$$

Example 2.6: Solve $(y + z)p + (z + x)q = x + y$

Solution: The given equation is of the form $Pp + Qq = R$.

Where, $P = y + z$, $Q = z + x$, and $R = x + y$

Now, the Langrage auxiliary equations are

$$\frac{dx}{P} = \frac{dy}{Q} = \frac{dz}{R}$$

$$\therefore \frac{dx}{y+z} = \frac{dy}{z+x} = \frac{dz}{x+y}$$

From each fraction, we can write

$$\therefore \frac{dx+dy+dz}{2(x+y+z)} = \frac{dx-dy}{y-x} = \frac{dy-dz}{z-y}$$

From the first two fractions, we have

$$\frac{dx+dy+dz}{2(x+y+z)} = -\frac{dx-dy}{x-y}$$

$$\therefore \frac{1}{2}\frac{d(x+y+z)}{(x+y+z)} = -\frac{d(x-y)}{x-y}$$

Taking integration on both the sides,

$$\therefore \frac{1}{2}\log(x+y+z) = -\log(x-y) + \log c_1$$

$$\therefore \log[(x+y+z)^{1/2}(x-y)] = \log c_1$$

$$\therefore (x+y+z)^{1/2}(x-y) = c_1$$

Hence, $u(x, y, z) = (x+y+z)^{1/2}(x-y) = c_1$

Now, by taking the last two fractions, we have

Taking integration on both the sides,

$$\log(x-y) = \log(y-z) + \log c_2$$

$$\therefore \frac{x-y}{y-z} = c_2$$

Hence, the complete solution is given by

$$F\left[(x + y + z)^{1/2}(x - y), \frac{x - y}{y - z}\right] = 0$$

Example 2.7: Solve $(z - y)p + (x - z)q = y - x$

Solution: The given equation is of the form $Pp + Qq = R$.

Where, $P = z - y$, $Q = x - z$, and $R = y - x$

Now, the Langrage auxiliary equations are

$$\frac{dx}{P} = \frac{dy}{Q} = \frac{dz}{R}$$

$$\frac{dx}{z - y} = \frac{dy}{x - z} = \frac{dz}{y - x}$$

Add each fraction and compare with the last fraction, we can write

$$\frac{dx + dy + dz}{z - y + x - z + y - x} = \frac{dz}{y - x}$$

$$\therefore \frac{dx + dy + dz}{0} = \frac{dz}{y - x}$$

$$\therefore dx + dy + dz = 0$$

Taking integration on both the sides

$$\therefore x + y + z = c_1$$

Hence, $u(x, y, z) = x + y + z = c_1$

For the second solution using x, y, and z as multipliers and comparing with the last fraction, we get

$$\frac{x\,dx + y\,dy + z\,dz}{zx - yx + xy - zy + yz - xz} = \frac{dz}{y - x}$$

$$\therefore \frac{x\,dx + y\,dy + z\,dz}{0} = \frac{dz}{y - x}$$

$$\therefore x\,dx + y\,dy + z\,dz = 0$$

Taking integration on both the sides

$$\therefore x^2 + y^2 + z^2 = c_2$$

Hence, $v(x, y, z) = x^2 + y^2 + z^2 = c_2$

Hence, the complete solution is given by

$$F[x + y + z, x^2 + y^2 + z^2] = 0$$

Example 2.8: Solve $x(y - z)p + y(z - x)q = z(x - y)$

Solution: The given equation is of the form $Pp + Qq = R$.

Where, $P = x(y - z)$, $Q = y(z - x)$, and $R = z(x - y)$

Now, the Langrage auxiliary equations are

$$\frac{dx}{P} = \frac{dy}{Q} = \frac{dz}{R}$$

$$\frac{dx}{x(y - z)} = \frac{dy}{y(z - x)} = \frac{dz}{z(x - y)}$$

Add each fraction and compare with the last fraction, we can write

$$\frac{dx + dy + dz}{x(y - z) + y(z - x) + z(x - y)} = \frac{dz}{z(x - y)}$$

$$\therefore \frac{dx + dy + dz}{0} = \frac{dz}{z(x - y)}$$

$$\therefore dx + dy + dz = 0$$

Taking integration on both the sides

$$\therefore x + y + z = c_1$$

Hence, $u(x, y, z) = x + y + z = c_1$

For the second solution using $\frac{1}{x}, \frac{1}{y}, \frac{1}{z}$ as multipliers and comparing with the last fraction, we get

$$\frac{\frac{dx}{x} + \frac{dy}{y} + \frac{dz}{z}}{(y - z) + (z - x) + (x - y)} = \frac{dz}{z(x - y)}$$

$$\therefore \frac{\frac{dx}{x} + \frac{dy}{y} + \frac{dz}{z}}{0} = \frac{dz}{z(x - y)}$$

$$\therefore \frac{dx}{x} + \frac{dy}{y} + \frac{dz}{z} = 0$$

Now, taking integration on both the sides, we have

$$\therefore \ \log x + \log y + \log z = \log c_2$$

$$\therefore \ \log(xyz) = \log c_2$$

$$\therefore xyz = c_2$$

Thus, $v(x, y, z) = xyz = c_2$

Hence, the complete solution is given by

$$F[x + y + z, xyz] = 0$$

Example 2.9: Solve $x^2(y - z)p + y^2(z - x)q = z^2(x - y)$

Solution: The given equation is of the form $Pp + Qq = R$.

Where, $P = x^2(y - z)$, $Q = y^2(z - x)$, and $R = z^2(x - y)$

Now, the Langrage auxiliary equations are

$$\frac{dx}{P} = \frac{dy}{Q} = \frac{dz}{R}$$

$$\therefore \quad \frac{dx}{x^2(y - z)} = \frac{dy}{y^2(z - x)} = \frac{dz}{z^2(x - y)}$$

For the first solution using $\frac{1}{x^2}, \frac{1}{y^2}, \frac{1}{z^2}$ as multipliers and comparing with the last fraction, we get

$$\therefore \quad \frac{\frac{dx}{x^2} + \frac{dy}{y^2} + \frac{dz}{z^2}}{(y - z) + (z - x) + (x - y)} = \frac{dz}{z^2(x - y)}$$

$$\therefore \quad \frac{\frac{dx}{x^2} + \frac{dy}{y^2} + \frac{dz}{z^2}}{0} = \frac{dz}{z^2(x - y)}$$

$$\therefore \quad \frac{dx}{x^2} + \frac{dy}{y^2} + \frac{dz}{z^2} = 0$$

Taking integration on both the sides, we get

$$-\frac{1}{x} - \frac{1}{y} - \frac{1}{z} = c_1{}'$$

$$\therefore \quad \frac{1}{x} + \frac{1}{y} + \frac{1}{z} = c_1$$

Hence, $u(x, y, z) = \frac{1}{x} + \frac{1}{y} + \frac{1}{z} = c_1$

For the second solution using $\frac{1}{x}, \frac{1}{y}, \frac{1}{z}$ as multipliers and comparing with the last fraction, we get

$$\frac{\frac{dx}{x} + \frac{dy}{y} + \frac{dz}{z}}{x(y-z) + y(z-x) + z(x-y)} = \frac{dz}{z^2(x-y)}$$

$$\therefore \frac{\frac{dx}{x} + \frac{dy}{y} + \frac{dz}{z}}{0} = \frac{dz}{z^2(x-y)}$$

$$\therefore \frac{dx}{x} + \frac{dy}{y} + \frac{dz}{z} = 0$$

Integrating both the sides, we get

$$\therefore \log x + \log y + \log z = \log c_2$$

$$\therefore \log(xyz) = \log c_2$$

$$\therefore xyz = c_2$$

Hence, $v(x, y, z) = xyz = c_2$

Therefore, the complete solution is given by

$$F\left[\frac{1}{x} + \frac{1}{y} + \frac{1}{z}, xyz\right] = 0$$

Example 2.10: Solve $(y^3x - 2x^4)\frac{\partial z}{\partial x} + (2y^4 - x^3y)\frac{\partial z}{\partial y} = 9z(x^3 - y^3)$

Solution: The given equation is of the form $Pp + Qq = R$.

Where, $P = (y^3x - 2x^4)$, $Q = (2y^4 - x^3y)$, and $R = 9z(x^3 - y^3)$

Now, the Langrage auxiliary equations are

$$\frac{dx}{P} = \frac{dy}{Q} = \frac{dz}{R}$$

$$\therefore \frac{dx}{y^3x - 2x^4} = \frac{dy}{2y^4 - x^3y} = \frac{dz}{9z(x^3 - y^3)}$$

For the first solution using $\frac{1}{x}$, $\frac{1}{y}$, $\frac{1}{3z}$ as multipliers and comparing with the last fraction, we get

$$\therefore \frac{\frac{dx}{x} + \frac{dy}{y} + \frac{dz}{3z}}{y^3 - 2x^3 + 2y^3 - x^3 + 3(x^3 - y^3)} = \frac{dz}{9z(x^3 - y^3)}$$

$$\therefore \frac{\frac{dx}{x} + \frac{dy}{y} + \frac{dz}{3z}}{0} = \frac{dz}{9z(x^3 - y^3)}$$

$$\therefore \frac{dx}{x} + \frac{dy}{y} + \frac{dz}{3z} = 0$$

Taking integration on both the sides, we get

$$\therefore xyz^{1/3} = c_1$$

Hence, $u(x, y, z) = xyz^{1/3} = c_1$

Now, for finding the second solution, consider the first two fractions,

$$\frac{dx}{y^3x - 2x^4} = \frac{dy}{2y^4 - x^3y}$$

$$(2y^4 - x^3y)dx + (2x^4 - y^3x)dy = 0$$

The above equation is a non-exact differential equation.

So, to convert that equation in the form of an exact differential equation, multiplying it with integrating factor $\frac{1}{x^3y^3}$, we get

$$\left(\frac{2y}{x^3} - \frac{1}{y^2}\right)dx + \left(\frac{2x}{y^3} - \frac{1}{x^2}\right)dy = 0$$

Here, $M = \left(\frac{2y}{x^3} - \frac{1}{y^2}\right)$, and $N = \left(\frac{2x}{y^3} - \frac{1}{x^2}\right)$

From the above equation, we can prove that $\frac{\partial M}{\partial y} = \frac{\partial N}{\partial x}$

Hence, for the exact differential equation, the general solution is

$$\int_{y \text{ is constant}} M \, dx + \int_{\text{terms free from } x} N \, dy = 0$$

$$\therefore \int_{y \text{ is constant}} \left(\frac{2y}{x^3} - \frac{1}{y^2} \right) dx + \int_{\text{terms free from } x} 0 \, dy = 0$$

$$\therefore \frac{-y}{x^2} - \frac{x}{y^2} = c_2$$

Hence, $v(x, y) = \frac{-y}{x^2} - \frac{x}{y^2} = c_2$

Therefore, the complete solution is given by

$$F \left[xyz^{1/3}, \frac{-y}{x^2} - \frac{x}{y^2} \right] = 0$$

2.2 NON-LINEAR FIRST-ORDER PARTIAL DIFFERENTIAL EQUATIONS

The partial differential equation in which either $\frac{\partial z}{\partial x}$ or $\frac{\partial z}{\partial y}$ or both appear in other than first degree or they multiply together or with the dependent variable (z) is called a non-linear partial differential equation of the first order. The general form of such an equation is $f(x, y, z, \frac{\partial z}{\partial x}, \frac{\partial z}{\partial y}) = 0$.

2.2.1 Charpit Method

To find the complete solution of the first-order non-linear partial differential equation of the form

$$f(x, y, z, p, q) = 0.$$

The main concept of Charpit method is the introduction of another first-order partial differential equation of the form

$$F(x, y, z, p, q) = 0.$$

Solve the above two equations for p and q, and substitute in

$$dz = p(x, y, z, a)dx + q(x, y, z, a)dy.$$

The solution of the above equation, if it exists, is the complete solution of the equation $f(x, y, z, p, q) = 0$.

Now, the main thing is to determine the $F(x, y, z, p, q) = 0$ which is compatible with the equation $dz = p(x, y, z, a)dx + q(x, y, z, a)dy$.

The necessary and sufficient condition is

$$\frac{\partial(f, F)}{\partial(x, p)} + p\frac{\partial(f, F)}{\partial(z, p)} + \frac{\partial(f, F)}{\partial(y, q)} + q\frac{\partial(f, F)}{\partial(z, q)} = 0$$

$$\therefore \left(\frac{\partial f}{\partial x}\frac{\partial F}{\partial p} - \frac{\partial f}{\partial p}\frac{\partial F}{\partial x}\right) + p\left(\frac{\partial f}{\partial z}\frac{\partial F}{\partial p} - \frac{\partial f}{\partial p}\frac{\partial F}{\partial z}\right) + \left(\frac{\partial f}{\partial y}\frac{\partial F}{\partial q} - \frac{\partial f}{\partial q}\frac{\partial F}{\partial y}\right)$$

$$+ q\left(\frac{\partial f}{\partial z}\frac{\partial F}{\partial q} - \frac{\partial f}{\partial q}\frac{\partial F}{\partial z}\right) = 0$$

$$\therefore f_p\frac{\partial F}{\partial x} + f_q\frac{\partial F}{\partial y} + (pf_p + qf_q)\frac{\partial F}{\partial z} - (f_x + pf_z)\frac{\partial F}{\partial p} - (f_y + qf_z)\frac{\partial F}{\partial q} = 0$$

The above equation is a linear partial differential equation.

So, the auxiliary equation is

$$\frac{dx}{f_p} = \frac{dy}{f_q} = \frac{dz}{pf_p + qf_q} = \frac{dp}{-(f_x + pf_z)} = \frac{dq}{-(f_y + qf_z)}.$$

The above equation is called the Charpit auxiliary equation.

The Charpit method is illustrated through the examples as follows.

Example 2.11: Solve $xp + yp = pq$

Solution: The given equation $f(x, y, z, p, q) = 0$ is

$$xp + yq - pq = 0$$

$$\Rightarrow f_x = p, \quad f_y = q, \quad f_z = 0, \quad f_p = x - q, \quad f_q = y - p.$$

The Charpit auxiliary equation is

$$\frac{dx}{f_p} = \frac{dy}{f_q} = \frac{dz}{pf_p + qf_q} = \frac{dp}{-(f_x + pf_z)} = \frac{dq}{-(f_y + qf_z)}$$

$$\Rightarrow \frac{dp}{p} = \frac{dq}{q} = \frac{dz}{-p(x-q)-q(y-p)} = \frac{dx}{-(x-q)} = \frac{dy}{-(y-p)}.$$

From the first two members of the above equation, we get

$$\int \frac{dp}{p} = \int \frac{dq}{q}$$

$$\therefore \ \log p = \log q + \log a$$

$$\therefore p = aq.$$

Substitute p in the equation $xp + yp - pq = 0$, we get

$$q(ax + y) = aq^2$$

$$\therefore q = \frac{y + ax}{a}$$

$$\therefore p = aq = y + ax.$$

Now, $dz = pdx + qdy$

$$\Rightarrow dz = (y + ax)dx + \frac{y + ax}{a}dy$$

$$\Rightarrow adz = (y + ax)adx + (y + ax)dy$$

$$\Rightarrow adz = (y + ax)(adx + dy).$$

On integrating, we get

$$az = \frac{(y + ax)^2}{2} + b, \text{ which is the required solution.}$$

Example 2.12: Solve $(p^2 + q^2)\, y = qz$

Solution: The given equation $f(x, y, z, p, q) = 0$ is

$$(p^2 + q^2)y - qz = 0$$

$$\Rightarrow f_x = 0, \quad f_y = p^2 + q^2, \quad f_z = -q, \quad f_p = 2py, \quad f_q = 2qy - z.$$

The Charpit auxiliary equation is

$$\frac{dx}{f_p} = \frac{dy}{f_q} = \frac{dz}{pf_p + qf_q} = \frac{dp}{-(f_x + pf_z)} = \frac{dq}{-(f_y + qf_z)}$$

$$\frac{dx}{2py} = \frac{dy}{2qy - z} = \frac{dz}{2p^2y + 2q^2y - qz} = \frac{dp}{pq} = \frac{dq}{-[(p^2 + q^2) - q^2]}.$$

From the last two fractions of the above equation, we have

$$\frac{dp}{pq} = \frac{dq}{-p^2}$$

$$\therefore \frac{dp}{q} = \frac{dq}{p}$$

$$\therefore pdp + qdq = 0.$$

Taking integration on both sides, we have

$$\therefore p^2 + q^2 = a.$$

Substitute $p^2 + q^2 = a$ in the equation $(p^2 + q^2) y - qz = 0$, we get

$$ay - qz = 0$$

$$\therefore q = \frac{ay}{z}$$

$$\therefore p = \sqrt{a - \left(\frac{ay}{z}\right)^2}.$$

Now, $dz = pdx + qdy$

$$\Rightarrow dz = \sqrt{a - \left(\frac{ay}{z}\right)^2} \, dx + \frac{ay}{z} dy$$

$$\Rightarrow dz = \frac{\sqrt{az^2 - a^2y^2}}{z}dx + \frac{ay}{z}dy$$

$$\Rightarrow zdz = \sqrt{az^2 - a^2y^2}\,dx + aydy$$

$$\Rightarrow zdz - aydy = \sqrt{az^2 - a^2y^2}\,dx$$

$$\Rightarrow \frac{zdz - aydy}{\sqrt{az^2 - a^2y^2}} = dx$$

$$\Rightarrow \frac{d\,(az^2 - a^2y^2)^{1/2}}{a} = dx.$$

Taking integration on both the sides, we have

$$\frac{(az^2 - a^2y^2)^{1/2}}{a} = x + b.$$

Hence, the required complete solution is

$$(x + b)^2 + y^2 = \frac{z^2}{a}.$$

2.2.2 Special Types of First-Order Partial Differential Equations

Type-1: Equations Involving Only p and q

If the partial differential equations contain only p and q and not x, y, z, i.e. $f(p, q) = 0$, then substitute $p = a$ and $q = b$ in $f(p, q) = 0$, where a and b are constants.
Therefore, $f(a, b) = 0 \Rightarrow b = \phi(a)$.
Now, the complete solution is given by $z = ax + by + c$.
Hence, the complete solution is $z = ax + \phi(a)\,y + c$; a and c are arbitrary constants.

Example 2.13: Solve $p + q = pq$

Solution: The given equation contains only p and q, i.e. it is of the form $f(p, q) = 0$.

Putting $p = a$ and $q = b$ in the given equation, we get

$$a + b = ab$$

$$\therefore a = ab - b$$

$$\therefore b = \frac{a}{a - 1}.$$

Now, the complete solution is given by

$$z = ax + by + c$$

$$\therefore z = ax + \frac{a}{a - 1} y + c$$

Example 2.14: Solve $\sqrt{p} + \sqrt{q} = 1$

Solution: The given equation contains only p and q, i.e. it is of the form $f(p, q) = 0$

Putting $p = a$ and $q = b$ in the given equation, we get

$$\sqrt{a} + \sqrt{b} = 1$$

$$\therefore b = (1 - \sqrt{a})^2.$$

Now, the complete solution is given by

$$z = ax + by + c$$

$$\therefore z = ax + (1 - \sqrt{a})^2 y + c.$$

Type-2: Equations Not Involving Independent Variables
    The partial differential equation contains only z, p, and q and not x and y, of the form $f(z, p, q) = 0$.
    Assume $p = aq$ or $q = ap$.
    Therefore, $f(z, p, q) = 0$ can be written as either in the form of $f(z, p) = 0$ or $f(z, q) = 0$.
    Then from $f(z, p) = 0$, derive $p = \phi(z)$ and then find $q = \psi(z)$ from $q = ap$.
    Or from $f(z, q) = 0$, derive $q = \phi(z)$, and then find $p = \psi(z)$ from $p = aq$.
    Substitute p and q in the differential relation $dz = pdx + qdy$.
    Now, by taking integration on both sides we can obtain the complete solution.

Example 2.15: Solve $q^2 = z^2 p^2 (1 - p^2)$

Solution: The given equation is of the form $f(z, p, q) = 0$.

Let, $q = ap$ substitute in the given equation, we have

$$a^2 p^2 = z^2 p^2 (1 - p^2)$$

$$\therefore a^2 = z^2 (1 - p^2)$$

$$\therefore p^2 = 1 - \frac{a^2}{z^2}$$

$$\therefore p = \frac{\sqrt{z^2 - a^2}}{z}.$$

So, from $q = ap \Rightarrow q = a\frac{\sqrt{z^2 - a^2}}{z}$.

Now, $dz = pdx + qdy$

$$\therefore dz = \frac{\sqrt{z^2 - a^2}}{z} dx + a\frac{\sqrt{z^2 - a^2}}{z} dy$$

$$\therefore \frac{z}{\sqrt{z^2 - a^2}} dz = dx + ady.$$

Take integration on both the sides, we have the complete solution

$$\frac{1}{2} \frac{\sqrt{z^2 - a^2}}{\frac{1}{2}} = x + ay + c$$

$$\therefore z^2 = a^2 + (x + ay + c)^2.$$

Example 2.16: Solve $p(1 + q) = qz$

Solution: The given equation is of the form $f(z, p, q) = 0$.

Let, $p = aq$ substitute in the given equation, we have

$$aq(1 + q) = qz$$

$$\therefore a(1 + q) = z$$

$$\therefore q = \frac{z}{a} - 1$$

$$\therefore q = \frac{z - a}{a}.$$

So, from $p = aq \Rightarrow p = a\frac{z-a}{a} \Rightarrow p = z - a$.

Now, $dz = pdx + qdy$

$$\therefore dz = (z - a)dx + \frac{(z - a)}{a}dy$$

$$\therefore \frac{dz}{(z - a)} = dx + \frac{1}{a}dy.$$

Take integration on both the sides, we have the complete solution

$$\log(z - a) = x + \frac{y}{a} + c.$$

Type-3: Separable Equations

Such equations are of the form $f(x, p) = g(y, q)$ equation in which z is absent and the terms containing x and p are separated from those containing y and q.
Let $f(x, p) = g(y, q) = a$.
Solving, $f(x, p) = a$ we obtain $p = F(x)$ and $g(y, q) = a$ gives $q = G(y)$.
Now, the differential relation is $dz = pdx + qdy$

$$\Rightarrow dz = F(x)dx + G(y)dy.$$

Integrating both the sides, we get $z = \int F(x)dx + \int G(y)dy + b$, which is the complete integral.

Example 2.17: Solve $p^2 - q^2 = x - y$

Solution: Given $p^2 - q^2 = x - y$

$$\Rightarrow p^2 - x = q^2 - y.$$

Which is of the form $f(x, p) = g(y, q)$

Let $p^2 - x = q^2 - y = a$

$$\therefore p = \sqrt{a + x} \text{ and } q = \sqrt{a + y}.$$

Putting these values in the equation

$$dz = pdx + qdy.$$

We obtain,

$$dz = \sqrt{a + x}\,dx + \sqrt{a + y}\,dy.$$

Integrating on both the sides, we get the complete solution

$$z = \frac{2}{3}(a + x)^{3/2} + \frac{2}{3}(a + y)^{3/2} + b.$$

Example 2.18: Solve $p^2 y(1 + x^2) = qx^2$

Solution: Given $p^2 y(1 + x^2) = qx^2$

$$\Rightarrow \frac{p^2(1 + x^2)}{x^2} = \frac{q}{y}.$$

Which is of the form $f(x, p) = g(y, q)$

Let, $\frac{p^2(1 + x^2)}{x^2} = \frac{q}{y} = a.$

$$\therefore p = \frac{x\sqrt{a}}{\sqrt{1 + x^2}}, \quad q = ay.$$

Putting these values in the equation

$$dz = pdx + qdy.$$

We obtain,

$$dz = \frac{x\sqrt{a}}{\sqrt{1 + x^2}} dx + ay\, dy$$

Integrating on both the sides, we get the complete solution

$$z = \sqrt{a}\sqrt{1 + x^2} + \frac{a}{2}y^2 + b.$$

Type-4: Clairaut's Form

A partial differential equation of the form $z = px + qy + f(p, q)$ is called Clairaut's equation.

Its complete solution is given by $z = px + qy + f(a, b)$, which can be verified by direct differentiation as $p = a$ and $q = b$.

Example 2.19: Solve $z = px + qy + \sqrt{1 + p^2 + q^2}$

Solution: The given equation is of the form $z = px + qy + f(p, q)$.

Putting $p = a$ and $q = b$

$z = ax + by + \sqrt{1 + a^2 + b^2}$ is the required complete solution.

Example 2.20: Solve $(p + q)(z - xp - yq) = 1$

Solution: The given partial differential equation can be rewritten as

$$z = xp + yq + \frac{1}{p + q}.$$

Which is of the form $z = px + qy + f(p, q)$

Putting $p = a$ and $q = b$

$$z = ax + by + \frac{1}{a + b}.$$

EXERCISES

Q1 Solve $(x^2 - y^2 - z^2)p + 2xyq = 2xz$

Q2 Solve $xp + yq = z$

Q3 Solve $y^2p - xyq = x(z - 2y)$

Q4 Solve $yp - xq + x^2 - y^2 = 0$

Q5 Solve $p\tan x + q\tan y = \tan z$

Q6 Solve $p\left(\frac{y-z}{yz}\right) + q\left(\frac{z-x}{zx}\right) = \frac{x-y}{xy}$

Q7 Solve $(z^2 - 2zy - y^2)p + (xy + xz)q = xy - xz$

Q8 Solve $p^2 + q^2 = npq$

Q9 Solve $3p^2 - 2q^2 = 4pq$

Q10 Solve $p^2z^2 + q^2 = 1$

Q11 Solve $p^2 + q^2 = z$

Q12 Solve $p^2 - x^2 = y^2 - q^2$

Q13 Solve $p + q = \sin x + \sin y$

Q14 Solve $z = px + qy - 2\sqrt{pq}$

Q15 Solve $pqz = p^2(xq + p^2) + q^2(yp + q^2)$

ANSWERS

1    $F\left(\frac{y}{z}, \frac{x^2+y^2+z^2}{z}\right) = 0$

2    $F\left(\frac{x}{y}, \frac{y}{z}\right) = 0$

3    $F(y^2 - yz, x^2 + y^2) = 0$

4    $F(x^2 + y^2, xy - z) = 0$

5    $F\left(\frac{\sin y}{\sin x}, \frac{\sin z}{\sin x}\right) = 0$

6    $F(y^2 + z^2, x^3 + y^3) = 0$

7    $F(x^2 + y^2 + z^2, y^2 - 2yz - z^2) = 0$

8    $z = ax + \frac{ay}{2}\left(n \pm \sqrt{n^2 - 4}\right) + c$

9    $z = a\left(x + \frac{-2 \pm \sqrt{10}}{2}y\right) + c$

10   $\left[z\sqrt{a^2 + z^2} + a^2\log\left(z + \sqrt{a^2 + z^2}\right)\right] = 2(x + ay + b)$

11   $4(1 + a^2)z = (x + ay + b)^2$

12   $z = \frac{2}{3}(x^2 + a)^{3/2} + \frac{2}{3}(y^2 - a)^{3/2} + b$

13   $z = a(x - y) - (\cos x + \cos y) + b$

14   $z = ax + by - 2\sqrt{ab}$

15   $z = ax + by + \frac{a^4 + b^4}{ab}$

3 Second- and Higher-Order Linear Partial Differential Equations

In the previous chapter, we studied first-order linear and non-linear partial differential equations with some solution methods. In this chapter, we shall discuss the concept of second- and higher-order linear partial differential equations. We shall define homogeneous kinds of partial differential equations that are linear with constant coefficients. We introduce two solutions namely complementary functions and particular integrals and their addition will become the complete solution of the given problem. We shall evaluate a variety of examples, so readers can easily understand the concept. In the next section, we shall discuss the classification of second-order partial differential equations with some illustrations. In the last section, we shall study the method of separation of variables, which is a very famous, strong, and useful method to solve second-order linear partial differential equations with some examples. Exercises are provided with answers so students can do the practice.

3.1 HOMOGENEOUS LINEAR PARTIAL DIFFERENTIAL EQUATIONS WITH CONSTANT COEFFICIENTS

An equation of the form

$$a_0 \frac{\partial^n z}{\partial x^n} + a_1 \frac{\partial^n z}{\partial x^{n-1}\partial y} + a_2 \frac{\partial^n z}{\partial x^{n-2}\partial y^2} + \ldots\ldots\ldots + a_n \frac{\partial^n z}{\partial y^n} = F(x, y). \quad (3.1)$$

Where $a_0, a_1, a_2, \ldots\ldots a_n$ are constants, is called a homogeneous linear partial differential equation of the nth order with constant coefficients.

Let $D = \frac{\partial}{\partial x}$ and $D' = \frac{\partial}{\partial y}$

∴ The above general equation becomes

$[a_0 D^n + a_1 D^{n-1}D' + \ldots\ldots + a_n (D')^n]z = F(x, y)$, which can be written in the form of

$$f(D, D')z = F(x, y).$$

Now, its complete solution is G.S. = C.F. + P.I.

Where the complementary function (C.F.) is the solution of the equation $f(D, D')z = 0$ and particular integral (P.I.) is a particular solution of equation (3.1).

Rules for finding C.F.

Putting $D = m$ and $D' = 1$ in $f(D, D')z = 0$, we get the auxiliary equation.

1. When the auxiliary equation has distinct roots $m_1, m_2, \ldots m_n$ then C.F. is
$$z = f_1(y + m_1x) + f_2(y + m_2x) + f_3(y + m_3x) + \ldots + f_n(y + m_nx).$$

2. When the auxiliary equation has equal roots i.e. $m_1 = m_2$ and distinct roots $m_3 \ldots m_n$ then the C.F. is
$$z = f_1(y + m_1x) + xf_2(y + m_1x) + f_3(y + m_3x) + \ldots + f_n(y + m_nx).$$

Rules for finding P.I.

By (3.1), $f(D, D')z = F(x, y)$

$$\therefore \text{P.I.} = \frac{1}{f(D, D')} F(x, y)$$

1. When $F(x, y) = e^{ax+by}$
$$\therefore \text{P.I.} = \frac{1}{f(D, D')} e^{ax+by}$$
$$= \frac{1}{f(a, b)} e^{ax+by}; \quad \text{(i. e. replace D by a and D' by b, provided } f(a, b) \neq 0)$$

2. When $F(x, y) = \sin(ax + by)[or \cos(ax + by)]$
$$\text{P.I.} = \frac{1}{f(D, D')} \sin(ax + by)[or \cos(ax + by)]$$
$$= \frac{1}{f(a, b)} \sin(ax + by)[or \cos(ax + by)]$$
(i. e. replace DD' by $-ab$ and D^2 by $-a^2$, D'^2 by $-b^2$)

3. When $F(x, y) = x^m y^n$
$$\text{P.I.} = \frac{1}{f(D, D')} x^m y^n = [f(D, D')]^{-1} x^m y^n$$
If $m < n$, expand $[f(D, D')]^{-1}$ in powers of $\frac{D}{D'}$
If $m > n$, expand $[f(D, D')]^{-1}$ in powers of $\frac{D'}{D}$

Remark
$$\frac{1}{D} F(x, y) = \int_{y \text{ constant}} F(x, y)\, dx \text{ and } \frac{1}{D'} F(x, y) = \int_{x \text{ constant}} F(x, y)\, dx$$

Example 3.1: Solve $(D^2 - 2DD' + D'^2)z = e^{x+4y}$

Solution: The auxiliary equation is $m^2 - 2m + 1 = 0$ (assuming $D = m$ and $D' = 1$)

$$\therefore (m - 1)^2 = 0$$
$$\therefore m = 1, 1$$
$$\text{Hence, C.F.} = (f_1(y + x) + xf_2(y + x))$$

$$\text{P.I.} = \frac{1}{(D^2 - 2DD' + D'^2)}e^{x+4y}$$

$$= \frac{1}{(D - D')^2}e^{x+4y} \qquad \text{(replace } D \text{ by 1 and } D' \text{ replace by 4)}$$

$$= \frac{1}{(1-4)^2}e^{x+4y}$$

$$\text{P.I.} = \frac{e^{x+4y}}{9}$$

$$z = (f_1(y + x) + xf_2(y + x)) + \frac{e^{x+4y}}{9}$$

Example 3.2: $(D^3 - 3D^2D' + 4D'^3)z = e^{x+2y}$

Solution: The auxiliary equation is

$$m^3 - 3m^2 + 4 = 0$$
$$\therefore m = -1, 2, 2$$
$$\text{C.F.} = (f_1(y - x) + f_2(y + 2x) + xf_3(y + 2x))$$

$$\text{P.I.} = \frac{1}{(D^3 - 3D^2D' + D'^3)}e^{x+2y}$$

$$= \frac{1}{(D - D')(D + 2D')^2}e^{x+2y}$$

$$= \frac{1}{(1-2)(1+4)^2}e^{x+2y}$$

$$= -\frac{e^{x+2y}}{25}$$

$$z = (f_1(y - x) + f_2(y + 2x) + xf_3(y + 2x)) - \frac{e^{x+2y}}{5}$$

Example 3.3: $(D^2 - 2DD')z = 2\cos(x + y)$

Solution: The auxiliary equation is

$$m^2 - 2m = 0$$
$$\therefore m = 0, 2$$
$$\text{C.F.} = (f_1(y) + f_2(y + 2x))$$

$$\text{P.I.} = \frac{1}{(D^2 - 2DD')}2\cos(x + y)$$

$$= \frac{1}{(-1 - 2(-1))}2\cos(x + y) = 2\cos(x + y)$$

The General Solution is $z = (f_1(y) + f_2(y + 2x)) + 2 \cos (x+y)$.

Example 3.4: $z_{xx} + 2z_{xy} + z_{yy} = x^2 + y^2$

Solution: The auxiliary equation is

$$m^2 + 2m + 1 = 0$$
$$\therefore \quad (m + 1)^2 = 0$$
$$\therefore \quad m = -1, -1$$
$$\therefore \quad \text{C.F.} = f(y - x) + xf(y - x)$$

$$z_{xx} + 2z_{xy} + z_{yy} = x^2 + y^2$$

$$= \frac{1}{D^2} [\frac{1}{1 + \frac{2D'}{D} + \frac{D'^2}{D^2}}](x^2 + y^2)$$

$$= \frac{1}{D^2} [1 + \frac{2D'}{D} + \frac{D'^2}{D^2}]^{-1}(x^2 + y^2)$$

$$= \frac{1}{D^2} [1 - (\frac{2D'}{D} + \frac{D'^2}{D^2}) + (\frac{2D'}{D} + \frac{D'^2}{D^2})^2 - \text{..........}](x^2 + y^2)$$

$$= \frac{1}{D^2} [x^2 + y^2 - \frac{2(2y)}{D} - \frac{1}{D^2}(2) + 4\frac{1}{D^2}(2)]$$

$$= \frac{1}{D^2} [x^2 + y^2 - 4xy - x^2 + 4x^2]$$

$$= \frac{x^4}{3} - \frac{2x^3y}{3} + \frac{x^2y^2}{2}.$$

Example 3.5: $(D^2 + 3DD' + 2D'^2)z = x + y$

Solution: The auxiliary equation is

$$m^2 + 3m + 2 = 0$$

$$m = -1, \; m = -2$$

$$\text{C.F.} = f_1(y - x) + f_2(y - 2x)$$

Now, $\text{P.I.} = \dfrac{1}{D^2 + 3DD' + 2D'^2}(x + y)$

$$= \dfrac{1}{D^2(1 + \frac{3D'}{D} + \frac{2D'^2}{D^2})}(x + y)$$

$$= \dfrac{1}{D^2}(1 - (\dfrac{3D'}{D} + \dfrac{2D'^2}{D^2}))(x + y)$$

$$= \dfrac{1}{D^2}(x + y - \dfrac{3}{D}(1))$$

$$= \dfrac{1}{D^2}\left(x + y - 3\int 1 \, dx\right)$$

$$= \dfrac{1}{D^2}(x + y - 3x)$$

$$= \iint (y - 2x)dxdx$$

$$= \dfrac{yx^2}{2} - \dfrac{x^3}{3}$$

Example 3.6: $(D^2 - DD' - 2D'^2)z = 2x + 3y + e^{3x+4y}$

Solution: The auxiliary equation is

$$m^2 - m - 2 = 0 \Rightarrow m = 2, -1$$

$$\text{C.F.} = f_1(y + 2x) + f_2(y - 2x)$$

Now, $\text{P.I.} = \dfrac{1}{(D^2 - DD' - 2D'^2)}[2x + 3y + e^{3x+4y}]$

$$\text{P.I.} = \frac{1}{(D^2 - DD' - 2D')}[2x + 3y] + \frac{1}{(D - 2D')(D + D')}e^{3x+4y}$$

$$= \frac{1}{D^2\left(1 - \left(\dfrac{D'}{D} + 2\dfrac{D'^2}{D^2}\right)\right)}(2x + 3y) + \frac{1}{(3-8)(3+4)}e^{3x+4y}$$

$$= \frac{1}{D^2}\left(1 + \left(\frac{D'}{D} + 2\frac{D'^2}{D^2}\right)\right)(2x + 3y) - \frac{1}{35}e^{3x+4y}$$

$$= \frac{5x^3}{6} + \frac{3x^2y}{2} - \frac{1}{35}e^{3x+4y}$$

The General Solution is

$$z = f_1(y + 2x) + f_2(y - x) + \frac{5x^3}{6} + \frac{3x^2y}{2} - \frac{1}{35}e^{3x+4y}.$$

3.2 CLASSIFICATION OF SECOND-ORDER LINEAR PARTIAL DIFFERENTIAL EQUATIONS

The second-order partial differential equation is of the form $f(x, y, u, u_x, u_y, u_{xx}, u_{xy}, u_{yy}) = 0$, where x and y are independent variables and u is the function of x and y.

The most general form of a linear, second-order partial differential equation in two independent variables x and y, and the dependent variable $u(x, y)$ is
$$Au_{xx} + Bu_{xy} + Cu_{yy} + Du_x + Eu_y + Fu + G = 0,$$ where A to G are constants.

$$\text{This equation is called} \begin{cases} \text{Elliptic for } B^2 - 4ac < 0 \\ \text{Parabolic for } B^2 - 4ac = 0 \\ \text{Hyperbolic for } B^2 - 4ac > 0 \end{cases}$$

Example 3.7: Classify the following differential equation $2\dfrac{\partial^2 u}{\partial x^2} + 4\dfrac{\partial^2 u}{\partial x \partial y} + 3\dfrac{\partial^2 u}{\partial y^2} = 0$

Solution

Here, $A = 2$, $B = 4$, $C = 3$.

Let,

$$B^2 - 4AC = 16 - 24$$
$$= -8 < 0.$$

∴ The equation is elliptic.

Example 3.8: Classify the following differential equation $4\frac{\partial^2 u}{\partial x^2} - 9\frac{\partial^2 u}{\partial x \partial y} + 5\frac{\partial^2 u}{\partial y^2} = 0$

Solution: Here, $A = 4$, $B = -9$, $C = 5$.

Let,

$$B^2 - 4AC = 81 - 80$$
$$= 1 > 0.$$

∴ The equation is hyperbolic.

Example 3.9: Classify the following differential equation $\frac{\partial^2 u}{\partial x^2} + 4\frac{\partial^2 u}{\partial x \partial y} + 4\frac{\partial^2 u}{\partial y^2} = 0$

Solution: Here, $A = 1$, $B = 4$, $C = 4$.

Let,

$$B^2 - 4AC = 16 - 16$$
$$= 0.$$

∴ The equation is parabolic.

3.3 METHOD OF SEPARATION OF VARIABLES

The method of separation of variables is a powerful method to solve the linear partial differential equation in this method, we assume the solution as a product of two functions, each of which involves only one independent variable, by substituting the product functions, and its partial derivatives in the given equation we determine each of the function. If the problem involves more than two independent variables, the product involves more functions.

Example 3.10: By the method of variable separation method, solve

$$3\frac{\partial u}{\partial x} = \frac{\partial u}{\partial t} + u, \quad \text{where } u(x, 0) = 5e^{-2x}.$$

Solution: Let us assume that the solution of the given equation is as

$$u(x, t) = F(x)G(t).$$

Then, we obtain

$$\frac{\partial u}{\partial x} = F'(x)G(t)\frac{\partial u}{\partial t} = F(x)G'(t).$$

Substituting these values in the given equation, we obtain

$$3F'(x)G(t) = F(x)G'(t) + F(x)G(t)$$

$$3F'(x)G(t) = F(x)[G'(t) + G(t)]$$

$$3\frac{F'(x)}{F(x)} = [\frac{G'(t)}{G(t)} + 1] \text{ for all } x \text{ and } t.$$

Since L.H.S. is a function of x only, R.H.S. is the function of t only, and x and t are independent variables, we obtain $3(\frac{F'(x)}{F(x)}) = k$

$$\therefore [\frac{G'(t)}{G(t)} + 1] = k \text{ for all } x \text{ and } t.$$

Integrating on both the sides, we obtain

$$\ln F(x) = \frac{k}{3}x + \ln c_1 \text{ and } \ln G(t) + t = kt + c_2'$$

$$\Rightarrow F(x) = c_1 e^{\frac{kx}{3}} \text{ and } \ln G(t) = t(k-1) + c_2'$$

$$\Rightarrow F(x) = c_1 e^{\frac{kx}{3}} \text{ and } G(t) = c_2 e^{t(k-1)} \text{ where } c_2 = e^{c/2}.$$

Putting these values in $u(x, t) = F(x)G(t)$, we obtain

$$u(x, t) = c_1 c_2 e^{\frac{kx}{3}} e^{t(k-1)}.$$

Using the condition $u(x, 0) = 5e^{-2x}$, we obtain

$$5e^{-2x} = c_1 c_2 e^{\frac{kx}{3}}$$

$$\Rightarrow c_1 c_2 = 5 \text{ and } \frac{k}{3} = -2$$

$$\Rightarrow c_1 c_2 = 5 \text{ and } k = -6.$$

Hence, the required solution is

$$u(x, t) = 5e^{\frac{-6x}{3}} e^{t(-6-1)}$$

$$\therefore u(x, t) = 5e^{-2x} e^{-7t}.$$

Example 3.11: Using the method of separation variables, solve $u_{xx} = 16u_y$

Solution: Let us assume that the solution of the given equation is as

$$u(x, y) = F(x)G(y).$$

Then, we obtain

$$\frac{\partial u}{\partial x} = F'(x)G(y), \quad \frac{\partial u}{\partial y} = F(x)G'(y)$$

$$\frac{\partial^2 u}{\partial x^2} = F''(x)G(y).$$

Substituting these values in the given equation, we obtain

$$F''(x)G(y) = 16F(x)G'(y)$$

$$\frac{F''(x)}{F(x)} = 16\frac{G'(y)}{G(y)} \text{ for all } x \text{ and } y.$$

Since L.H.S. is a function of x only, R.H.S. is the function of y only, and x and y are independent variables, we obtain $\left(\frac{F''(x)}{F(x)}\right) = k^2$

$$\therefore 16\frac{G'(y)}{G(y)} = k^2 \text{ for all } x \text{ and } y.$$

We have the following three cases:

Case 1: $k^2 = 0$ then, we obtain
$F''(x) = 0$ and $G'(y) = 0$.
Integrating, we obtain $F(x) = (ax + b)$ and $G(y) = c$.
Hence, $u(x, y) = (ax + b)c = Ax + B$, where $A = ac$, $B = bc$ are arbitrary constants.

Case 2: When the constant is k^2, we obtain

$$\frac{F''(x)}{F(x)} = 16k^2$$

$$\therefore F''(x) = 16k^2 F(x)$$

$$\therefore F''(x) - 16k^2 F(x) = 0$$

$$\therefore (D^2 - 16k^2)F = 0.$$

The auxiliary equation is

$$m^2 - 16k^2 = 0$$

$$\therefore (m - 4k)(m + 4k) = 0$$

$$\therefore m = 4k, -4k.$$

The general solution is

$$F(x) = c_1 e^{4kx} + c_2 e^{-4kx}.$$

Solving the second equation, we obtain

$$\frac{G'(y)}{G(y)} = k^2$$

$$\therefore \ln G(y) = k^2 y + \ln c_3$$

$$\therefore G(y) = c_3 e^{k^2 y}.$$

Hence, we obtain
$u(x, y) = (Ae^{4kx} + Be^{-4kx})e^{ky}$, where A and B are arbitrary constants.

Case 3: When the constant is $- k^2$

$$\frac{F''(x)}{F(x)} = -16k^2$$

$$\therefore F''(x) = -16k^2 F(x)$$

$$\therefore F''(x) + 16k^2 F(x) = 0$$

$$\therefore (D^2 + 16k^2)F = 0.$$

The auxiliary equation is

$$m^2 + 16k^2 = 0$$

$$\therefore m = \pm 4ki.$$

The general solution is

$$F(x) = c_1 \cos 4kx + c_2 \sin 4kx.$$

Solving the second equation, we obtain

$$\frac{G'(y)}{G(y)} = -k^2$$

$$\therefore \ln G(y) = -k^2 y + \ln c_3$$

$$\therefore G(y) = c_3 e^{-k^2 y}.$$

Hence, we obtain

$$u(x, y) = (c_1 \cos 4kx + c_2 \sin 4kx)c_3 e^{-k^2 y}$$

$$\therefore u(x, y) = (A \cos 4kx + B \sin 4kx)e^{-k^2 y},$$

where A and B are arbitrary constants.

EXERCISES

Q1 Solve $(D^2 + 10DD' + 25D'^2) = e^{3x+2y}$

Q2 Solve $(D^2 + DD' - 6D'^2)z = y\cos x$

Q3 Solve $(D^2 + 2DD' + D'^2) = x^2 y$

Q4 Solve $\frac{\partial^2 z}{\partial x^2} - \frac{\partial^2 z}{\partial y^2} = x - y$

Q5 Solve $\frac{\partial^4 z}{\partial x^4} + 2\frac{\partial^4 z}{\partial x^2 \partial y^2} + \frac{\partial^4 z}{\partial y^4} = 36(x^2 + y^2)$

Q6 $(D^2 - DD')z = \cos x \cos 2y$

Q7 Solve $(D^3 + D^2 D' - DD'^2 - D'^3)z = e^x \cos 2y$

Q8 Solve $\frac{\partial^2 z}{\partial x^2} - 4\frac{\partial^2 z}{\partial y^2} = 3x - 4y$

Q9 Find the particular integral of the equation $(D^2 - D')z = 2y - x^2$

Q10 Solve $\frac{\partial^2 z}{\partial x^2} - 9\frac{\partial^2 z}{\partial y^2} = xy$

Q11 By the method of variable separation method, solve $\frac{\partial u}{\partial x} = 4\frac{\partial u}{\partial y}$, where $u(0, y) = 8e^{-3y}$

Q12 By the method of variable separation method, solve $3\frac{\partial u}{\partial x} + 2\frac{\partial u}{\partial y} = 0$, where $u(x, 0) = 4e^{-x}$

Q13 Solve $2\frac{\partial u}{\partial x} + \frac{\partial u}{\partial t} + 4u = 0$ using method of separation of variables, subject to the condition $u(x, 0) = 3e^{-x}(2 + e^x - e^{2x})$

Q14 Solve $u_x = 4u_t + u$. Given that $u(0, t) = 7e^{3t}$

Q15 Obtain the solution of the partial differential equation $\frac{\partial u}{\partial x} + 5\frac{\partial u}{\partial t} + u = 0$ with the condition $u(x, 0) = 3e^{-4x} + 2$

Q16 Solve $\frac{\partial^2 u}{\partial x^2} = \frac{\partial u}{\partial y} + 2u$ given $\frac{\partial u}{\partial x}(0, y) = 1 + e^{-3y}$

ANSWERS

1 $z = (f_1(y - 5x) + xf_2(y - 5x)) + \frac{e^{3x+2y}}{169}$

2 $u(x, t) = 2e^{-x+\frac{t}{2}}$

3 $z(x, t) = [Ae^{(1+\sqrt{1+k})x} + Be^{(1-\sqrt{1+k})x}]e^{-ky}$

4 $z = f_1(y + x) + f_2(y - 2x) + \frac{x^2}{6}(x - 3y)$

5 $z = f_1(y + ix) + xf_2(y + ix) + f_3(y - ix) + xf_4(y - ix) + 3\frac{x^2 y^2}{2} - \frac{x^6}{10}$

6 $z = f_1(y) + f_2(y + x) + \frac{1}{2}\cos(x + 2y) - \frac{1}{6}\cos(x - 2y)$

7 $z = f_1(y - x) + xf_2(y - x) + f_3(y + x) + \frac{e^x}{25}(\cos 2y + 2\sin 2y)$

8 $z = f_1(y + 2x) + f_2(y - 2x) + \frac{x^3}{2} - 2x^2 y$

9 particular integral $= x^2 y$

10 $z = f_1(y + 3x) + f_2(y - 3x) + \frac{x^3 y}{6} + \frac{x^4}{24}$

11 $u(x, y) = 8e^{-12-3y}$

12 $u(x, y) = 3e^{-5x-3y} + 2e^{-3x-2y}$

13 $u(x, y) = 3e^{-x}(2e^{-bt} + e^{x-4t} - e^{2x-2t})$

14 $u(x, t) = 7e^{13x+3t}$

15 $u(x, y) = 3e^{-4x+\frac{3}{5}t} + 2e^{\frac{-1}{5}t}$

16 $u(x, y) = \frac{1}{\sqrt{2}}\sinh\sqrt{2}x + e^{-3y}\sin x$

4 Applications of Partial Differential Equations

In the previous chapter, we introduced the method of separation of variables to solve second-order linear partial differential equations. In this chapter, once again we will use the method of separation of variables to find solutions to some important real-life problems. In the first section, we shall study the one-dimensional wave equation. In this section to solve the wave equation, we shall introduce the separation of variables method as examples. We shall explain D'Alemberts' Solution of the Wave Equation and Duhamel's Principle. In the next section, we shall study one-dimensional heat equation with some problems. In the last section, we shall discuss the Laplace equation with two examples in which Laplacian in cylindrical and spherical coordinates are studied. Lastly, some practice examples are given in the exercises.

4.1 ONE-DIMENSIONAL WAVE EQUATION

The equation $\frac{\partial^2 u}{\partial t^2} = c^2 \frac{\partial^2 u}{\partial x^2}$ governs the motion of the vibrating string over time, which is called the **one-dimensional wave equation**. It is a second-order partial differential equation, and it is linear and homogeneous.

4.1.1 THE SOLUTION OF THE WAVE EQUATION BY SEPARATION OF VARIABLES

There are several approaches to solving the wave equation. We will solve the wave equation using a technique called separation of variables, again, demonstrates one of the most widely used solution techniques for partial differential equations. The idea behind is to split up the original partial differential equation into a series of simpler ODEs, each of which we should be able to solve readily using tricks already learned.

The string is tightened at both the ends so displacement at $x = 0$ and $x = l$. So, boundary conditions are $u(0, t) = u(l, t) = 0$ for all values of t. Because it must be the case that the vertical displacement at the endpoints is 0 since they don't move up and down.

You might also note that we probably need to specify what the shape of the string is right when time $t = 0$, and you're right – to come up with a particular solution function, we would need to know $u(x, 0)$. Let the $u(x, 0) = f(x)$, we would also need to know the initial velocity of the string, which is just $u_t(x, 0) = g(t)$. These two requirements are called the **initial conditions** for the wave equation and are also necessary to specify a particular vibrating string solution.

To start the separation of variables technique, we make the key assumption that whatever the solution function is, that it can be written as the product of two independent functions, each one of which depends on just one of the two variables, x or t.

Thus, imagine that the solution function $u(x, t)$ can be written as

$$u(x, t) = F(x)G(t)$$

where F and G are single variable functions of x and t, respectively. Differentiating this equation $u(x, t)$ for twice with respect to each variable yields

$$\frac{\partial^2 u}{\partial x^2} = F''(x)G(t) \text{ and } \frac{\partial^2 u}{\partial x^2} = F(x)G''(t).$$

Thus, when we substitute these two equations back into the original wave equation

$$\frac{\partial^2 u}{\partial t^2} = c^2 \frac{\partial^2 u}{\partial x^2},$$

then, we get $\frac{\partial^2 u}{\partial t^2} = F(x)G''(t) = c^2 \frac{\partial^2 u}{\partial x^2} = c^2 F''(x)G(t)$

$$\therefore \frac{G''(t)}{c^2 G(t)} = \frac{F''(x)}{F(x)}.$$

Now, we have equality where the left-hand side just depends on the variable t, and the right-hand side just depends on x. Here comes the critical observation – how can two functions, one just depending on t, and one just on x, be equal for all possible values of t and x? The answer is that they must each be constant, for otherwise, the equality could not possibly hold for all possible combinations of t and x. Thus, we have

$$\therefore \frac{G''(t)}{c^2 G(t)} = \frac{F''(x)}{F(x)} = k$$

where k is a constant. First, let's examine the possible causes for k.

Case 1: $k = 0$

Then the above equations can be rewritten as

$$G''(t) = 0 \text{ and } F''(x) = 0$$

yielding with very little effort two solution functions for F and G:

$$G(t) = at + b \text{ and } F(x) = px + r$$

where a, b, p, and r are constants (note how easy it is to solve such simple ODEs versus trying to deal with two variables at once, hence the power of the separation of variables approach).

$$u(x, t) = F(x)G(t)$$

$$u(x, t) = (px + r)(at + b).$$

Now, we will apply the boundary conditions

$$\therefore u(0, t) = r(at + b) = 0$$

$$\therefore r = 0.$$

Hence, $u(x, t) = (px)(at + b)$.
  Again, $u(l, t) = pl(at + b) = 0$.
  But, $pl \neq 0$; Hence, $(at + b) = 0$.
  So, we have a trivial solution $u(x, t) = 0$ (which gives us the very dull solution equivalent to a flat, unplucked string).
  Thus, it must be the case that $u(x, t) = 0$, and we end up with the dull solution again, the only possible solution if we start with $k = 0$.

Case 2: $k > 0$
If k is positive. Let $k = \omega^2$, then from the equation we again start with
$G''(t) = \omega^2 c^2 G(t)$.
The auxiliary equation is

$$\therefore (D^2 - \omega^2 c^2) = 0$$

$$\therefore D = \pm \omega c$$

$$G(t) = ae^{\omega c t} + be^{-\omega c t}.$$

and for $F''(x) = \omega^2 F(x)$

$$\therefore (D^2 - \omega^2) = 0$$

$$\therefore D = \pm \omega$$

$$F(x) = pe^{\omega x} + qe^{-\omega x}.$$

Hence, we have $u(x, t) = (ae^{\omega c t} + be^{-\omega c t})(pe^{\omega x} + qe^{-\omega x})$.
  Now, apply $u(0, t) = 0$, we get

$$u(0, t) = (ae^{\omega ct} + be^{-\omega ct})(p + q) = 0.$$

But $(ae^{\omega ct} + be^{-\omega ct}) \neq 0$

hence, $p + q = 0 \Rightarrow p = -q$

$$u(x, t) = (ae^{\omega ct} + be^{-\omega ct})(pe^{\omega x} - pe^{-\omega x}).$$

Again apply $u(l, t) = 0$, we have

$$u(l, t) = (ae^{\omega ct} + be^{-\omega ct})(pe^{\omega l} - pe^{-\omega l}) = 0.$$

But $(e^{\omega l} - e^{-\omega l}) \neq 0$ and $(ae^{\omega ct} + be^{-\omega ct}) \neq 0$

$$\therefore p = 0.$$

Hence, $u(x, t) = 0$

Which is again a dull solution.

Case 3: $k < 0$

Let, $k = -\omega^2$

So, again we have

$$G''(t) = -\omega^2 c^2 G(t).$$

The auxiliary equation is

$$\therefore (D^2 + \omega^2 c^2) = 0$$

$$\therefore D = \pm \omega ci$$

$$G(t) = a \cos \omega ct + b \sin \omega ct$$

and for $F''(x) = -\omega^2 F(x)$

$$\therefore (D^2 + \omega^2) = 0$$

$$\therefore D = \pm \omega i$$

$$F(x) = p \cos \omega x + q \sin \omega x.$$

Hence, we have $u(x, t) = (a \cos \omega ct + b \sin \omega ct)(p \cos \omega x + q \sin \omega x).$

Now, apply $u(0, t) = 0$,

we get

$$u(0, t) = (a\cos\omega ct + b\sin\omega ct)(p\cos 0 + q\sin 0)$$

$$\therefore p = 0.$$

The solution reduced into the form

$$u(x, t) = (a\cos\omega ct + b\sin\omega ct)(q\sin\omega x).$$

Again apply $u(l, t) = 0$
we have

$$u(l, t) = (a\cos\omega ct + b\sin\omega ct)(q\sin\omega l) = 0$$

$$q\sin\omega l = 0 \text{ but } q \neq 0 \Rightarrow \sin\omega l = 0.$$

This means that $\omega l = n\pi$, where $n \in z$.

Here, n is the number of loops in the given string. We get

$$u(x, t) = \sin\left(\frac{n\pi x}{l}\right)\left\{A\cos\left(\frac{n\pi ct}{l}\right) + B\sin\left(\frac{n\pi ct}{l}\right)\right\} \text{ where, } A = aq \text{ and } B = bq.$$

This means we have infinitely many solutions, so by superposition theorem

$$u(x, t) = \sum_{n=1}^{\infty}\left\{A_n\cos\left(\frac{n\pi ct}{l}\right) + B_n\sin\left(\frac{n\pi ct}{l}\right)\right\}\sin\left(\frac{n\pi x}{l}\right).$$

Now, use initial conditions $u(x, 0) = f(x)$

$$f(x) = \sum_{n=1}^{\infty}\left\{A_n\sin\left(\frac{n\pi x}{l}\right)\right\}.$$

Which is the Fourier Sine series and A_n is the Fourier coefficient,

$$A_n = \frac{2}{l}\int_0^l f(x)\sin\left(\frac{n\pi x}{l}\right)dx.$$

where differentiate equation with respect to 't', we get

$$u_t(x, t) = \sum_{n=1}^{\infty} \left\{ -\frac{n\pi c}{l} A_n \sin\left(\frac{n\pi ct}{l}\right) + \frac{n\pi c}{l} B_n \cos\left(\frac{n\pi ct}{l}\right) \right\} \sin\left(\frac{n\pi x}{l}\right).$$

Now use another initial condition $u_t(x, 0) = g(t)$

$$g(t) = \sum_{n=1}^{\infty} \frac{n\pi c}{l} \left\{ B_n \sin\left(\frac{n\pi x}{l}\right) \right\},$$

which is the Fourier Sine series and B_n is the Fourier coefficient.

$$\frac{n\pi c}{l} B_n = \frac{2}{l} \int_0^l g(x) \sin\left(\frac{n\pi x}{l}\right) dx$$

$$B_n = \frac{2}{n\pi c} \int_0^l g(x) \sin\left(\frac{n\pi x}{l}\right) dx.$$

Example 4.1: A string of length $L = \pi$ has its ends fixed at $x = 0$ and $x = \pi$. At times $t = 0$ the string is given a shape defined by $f(x) = 50x(\pi - x)m$, and then it is released. Find the deflection of the string at any time t.

Solution: Let, $u(x, t)$ be the deflection of a string at any time t. Then $u(x, y)$ satisfies the wave equation.

$$\frac{\partial^2 u}{\partial^2 t^2} = c^2 \frac{\partial^2 u}{\partial^2 x^2}.$$

Its solution is given by

$$u(x, t) = \sum_{n=1}^{\infty} \left\{ A_n \cos\left(\frac{n\pi ct}{l}\right) + B_n \sin\left(\frac{n\pi ct}{l}\right) \right\} \sin\left(\frac{n\pi x}{l}\right)$$

where $A_n = \frac{2}{l} \int_0^l f(x) \sin\left(\frac{n\pi x}{l}\right) dx$ and $B_n = \frac{2}{n\pi c} \int_0^l g(x) \sin\left(\frac{n\pi x}{l}\right) dx$ ($n \in N$).

Since, the initial velocity $\frac{\partial u}{\partial t}(x, 0) = g(x) = 0$.

We have $B_n = 0$

Therefore, the solution

$$u(x, t) = \sum_{n=1}^{\infty} \left\{ A_n \cos\left(\frac{n\pi ct}{l}\right) + B_n \sin\left(\frac{n\pi ct}{l}\right) \right\} \sin\left(\frac{n\pi x}{l}\right)$$

becomes

$$u(x,\, t) = \sum_{n=1}^{\infty} \left\{ A_n \cos\left(\frac{n\pi ct}{l}\right) \right\} \sin\left(\frac{n\pi x}{l}\right).$$

Now, $A_n = \frac{2}{L} \int_0^L f(x) \sin\left(\frac{n\pi x}{L}\right) dx$

$$= \frac{2}{\pi} \int_0^\pi 50x\,(\pi - x) \sin(nx)\, dx$$

$$= \frac{100}{\pi} \int_0^\pi (\pi x - x^2) \sin(nx)\, dx$$

$$= \frac{100}{\pi} \left[(\pi x - x^2)\frac{-\cos(nx)}{n} - (\pi - 2x)\frac{-\sin(nx)}{n^2} + (-2)\frac{\cos(nx)}{n^3} \right]_0^\pi$$

$$= \frac{100}{\pi} \left[\left(0 - 0 - 2\frac{\cos(n\pi)}{n^3} \right) - \left(0 - 0 - 2\frac{\cos(0)}{n^3} \right) \right]$$

$$= \frac{200}{\pi n^3} [1 - (-1)^n].$$

Thus,

$$A_n = \begin{cases} \dfrac{400}{\pi n^3} & \text{if } n \text{ is odd} \\ 0 & \text{if } n \text{ is even} \end{cases}.$$

Hence, the required solution is

$$u(x,\, t) = \sum_{n=\text{odd}}^{\infty} \left\{ \frac{400}{\pi n^3} \cos\left(\frac{n\pi ct}{l}\right) \right\} \sin\left(\frac{n\pi x}{l}\right).$$

Example 4.2: If the string of length L is initially at rest in equilibrium position and each of its points is given the velocity,

$$\frac{\partial u}{\partial t}(x, 0) = \sin\left(\frac{3\pi x}{L}\right)\cos\left(\frac{2\pi x}{L}\right) \quad (0 \le x \le L \text{ at } t = 0).$$

Determine the displacement $u(x, t)$

Solution: The $u(x,t)$ satisfies the wave equation.

$$\frac{\partial^2 u}{\partial t^2} = c^2 \frac{\partial^2 u}{\partial x^2}.$$

Its solution is given by

$$u(x, t) = \sum_{n=1}^{\infty} \left\{ A_n \cos\left(\frac{n\pi ct}{l}\right) + B_n \sin\left(\frac{n\pi ct}{l}\right) \right\} \sin\left(\frac{n\pi x}{l}\right)$$

where $A_n = \frac{2}{l} \int_0^l f(x)\sin\left(\frac{n\pi x}{l}\right)dx$ and $B_n = \frac{2}{n\pi c} \int_0^l g(x)\sin\left(\frac{n\pi x}{l}\right)dx \ (n \in N)$.

Since, the initial velocity $u(x, 0) = f(x) = 0$ we have $A_n = 0$.

Therefore, the solution

$$u(x, t) = \sum_{n=1}^{\infty} \left\{ A_n \cos\left(\frac{n\pi ct}{l}\right) + B_n \sin\left(\frac{n\pi ct}{l}\right) \right\} \sin\left(\frac{n\pi x}{l}\right)$$

becomes

$$u(x, t) = \sum_{n=1}^{\infty} \left\{ B_n \sin\left(\frac{n\pi ct}{l}\right) \right\} \sin\left(\frac{n\pi x}{l}\right).$$

Also, the initial velocity is

$$\frac{\partial u}{\partial t}(x, 0) = \sin\left(\frac{3\pi x}{L}\right)\cos\left(\frac{2\pi x}{L}\right).$$

Differentiating the solution, we get

$$\frac{\partial u}{\partial t} = \sum_{n=1}^{\infty} \left\{ \left(\frac{n\pi c}{l}\right) B_n \left(\cos\left(\frac{n\pi ct}{l}\right)\right) \right\} \sin\left(\frac{n\pi x}{l}\right)$$

$$\frac{\partial u}{\partial t}(x, 0) = \sum_{n=1}^{\infty} \left\{ \left(\frac{n\pi c}{l}\right) B_n \right\} \sin\left(\frac{n\pi x}{l}\right) = \frac{1}{2}\left[\sin\left(\frac{5\pi x}{L}\right) + \sin\left(\frac{\pi x}{L}\right) \right]$$

$$\Rightarrow \sum_{n=1}^{\infty} \left\{ \left(\frac{n\pi c}{l}\right) B_n \right\} \sin\left(\frac{n\pi x}{l}\right) = \frac{1}{2}\left[\sin\left(\frac{5\pi x}{L}\right) + \sin\left(\frac{\pi x}{L}\right) \right].$$

Thus, we have

$$B_1 \frac{\pi c}{L} \sin\left(\frac{\pi x}{l}\right) + \ldots\ldots + B_5 \frac{5\pi c}{L} \sin\left(\frac{5\pi x}{l}\right) + \ldots\ldots = \left[\frac{1}{2} \sin\left(\frac{5\pi x}{L}\right) + \frac{1}{2} \sin\left(\frac{\pi x}{L}\right)\right].$$

Comparing the corresponding coefficients, we get

$$B_1 \frac{\pi c}{L} = \frac{1}{2}, \quad B_5 \frac{5\pi c}{L} = \frac{1}{2}$$

$$\therefore B_1 = \frac{L}{2\pi c}, \quad B_5 = \frac{L}{10\pi c}.$$

And the remaining coefficients are zero. Substituting these values of B_n in the solution, we get

$$u(x, t) = \frac{L}{2\pi c} \sin\left(\frac{\pi x}{L}\right) \sin\left(\frac{\pi c t}{L}\right) + \frac{L}{10\pi c} \sin\left(\frac{5\pi x}{L}\right) \sin\left(\frac{5\pi c t}{L}\right).$$

4.1.2 D'ALEMBERTS' SOLUTION OF THE WAVE EQUATION

$$\frac{\partial^2 u}{\partial t^2} = c^2 \frac{\partial^2 u}{\partial X^2}, \quad where \ c^2 = \frac{T}{\rho}$$

can be immediately obtained by suitably transforming above equation, namely, by introducing the new independent variables, $v = x + ct$, $z = x - ct$.

Then u becomes a function of v and z. The derivative in $\frac{\partial^2 u}{\partial t^2} = c^2 \frac{\partial^2 u}{\partial x^2}$, where $c^2 = \frac{T}{\rho}$ can now be expressed in terms of derivatives with respect to v and z by the use of the chain rule.

We see from $v = x + ct$, $z = x - ct$ that $v_x = 1$, $z_x = 1$.

For simplicity let us denote $u(x, t)$, as a function of v and z.

Then, $u_x = u_v v_x + u_z z_x = u_v + u_z$.

We now apply the chain rule to the right side. We assume that all the partial derivatives involved are continuous, so that, $u_{vz} = u_{zv}$.

Since $v_x = 1$, $z_x = 1$, we obtain

$$u_{xx} = (u_v + u_z)x = u_{vv} + 2u_{vz} + u_{zz}.$$

We transform the other derivative in $\frac{\partial^2 u}{\partial t^2} = c^2 \frac{\partial^2 u}{\partial x^2}$, Where $c^2 = \frac{T}{\rho}$ by the same process, finding

$$u_{tt} = C^2(u_{vv} - 2u_{vz} + u_{zz}).$$

By inserting these two results in $\frac{\partial^2 u}{\partial t^2} = c^2 \frac{\partial^2 u}{\partial x^2}$, where $c^2 = \frac{T}{\rho}$, we get

$$u_{vz} = \frac{\partial^2 u}{\partial z\, \partial v} = 0.$$

The point of the present method is that the resulting above equation can be readily solved by two successive integrations. Integrating u_{vz} with respect to z, we find

$$\frac{\partial u}{\partial v} = h(v)$$

where $h(v)$ is an arbitrary function of v. Integrating this with respect to v gives

$$u = \int h(v)\, dv + \psi(z),$$

where $\psi(z)$ is an arbitrary function of z. Since the integral is a function of v, say, $\phi(v)$ the solution u is of the form $u = \varphi(v) + \psi(z)$. Because of $v = x + ct$, $z = x - ct$,

$$u(x, t) = \varphi(x + ct) + \psi(x - ct).$$

This is known as **D'Alemberts' solution** of the wave equation $\frac{\partial^2 u}{\partial t^2} = c^2 \frac{\partial^2 u}{\partial x^2}$.

D'Alemberts' Solution Satisfying the Initial Condition

$$u_t(x, 0) = f(x)$$

$$u_t(x, 0) = g(x)$$

Now, by differentiating $u(x, t) = \varphi(x + ct) + \psi(x - ct)$ we have

$$u_t(x, t) = c\varphi'(x + ct) - c\psi'(x - ct).$$

Where primes denote derivatives with respect to the entire arguments $x + ct$ *and* $x - ct$, respectively.

From $u(x, t) = u_t(x, t)$, we have

$$u(x, 0) = \varphi(x) + \psi(x) = f(x), \tag{4.1}$$

$$u_t(x, 0) = c\varphi'(x) - c\psi'(x) = g(x). \tag{4.2}$$

Dividing $u_t(x, 0)$ by c and integrating with respect to x, we obtain

$$\varphi(x) - \psi(x) = \varphi(x_0) - \psi(x_0) + \frac{1}{c}\int_{x_0}^{x} g(s)ds. \tag{4.3}$$

If we add this to $u(x, 0)$, then ψ drops out and division by 2 gives

$$\varphi(x) = \frac{1}{2}f(x) + \frac{1}{2}\varphi(x_0) - \frac{1}{2}\psi(x_0) + \frac{1}{2c}\int_{x_0}^{x} g(s)ds. \tag{4.4}$$

Similarly, subtraction of equation (4.3) from equation (4.1) and division by 2 gives

$$\psi(x) = \frac{1}{2}f(x) - \frac{1}{2}\varphi(x_0) + \frac{1}{2}\psi(x_0) - \frac{1}{2c}\int_{x_0}^{x} g(s)ds. \tag{4.5}$$

In equation (4.4), we replace x by $x + ct$; we can get an integral form x_0 to $x + ct$. In equation (4.5), we replace x by $x - ct$

$$u(x, t) = \frac{1}{2}[f(x + ct) + f(x - ct)] + \frac{1}{2c}\int_{x-ct}^{x+ct} g(s)ds.$$

If the initial velocity is zero, we see that this reduces to

$$u(x, t) = \frac{1}{2}[f(x + ct) + f(x - ct)].$$

4.1.3 DUHAMEL'S PRINCIPLE FOR THE ONE-DIMENSIONAL WAVE EQUATION

Duhamel's Principle is the method to obtain solutions to non-homogeneous linear evolution equations like the wave equation, heat equation, and vibrating plate equation.

Application of Duhamel's Principle: Finite String Problem
If $U(x, t, s)$ is the solution to the problem

$$U_{tt} - c^2 U_{xx} = 0, \quad (x, t) \in (0, L) \times (0, \infty)$$

with boundary conditions $U(0, t, s) = 0$, $U(L, t, s) = 0$, $t > 0$, $s > 0$ and with initial conditions $U(x, 0, s) = 0$, $U_t(x, 0, s) = f(x, s)$, $s > 0$, then $u(x, t)$ defined by
$u(x, t) = \int_0^t U(x, t - \theta, \theta)\, d\theta$ is the solution to the non-homogeneous problem

$$u_{tt} - c^2 u_{xx} = f(x, t), \quad (x, t) \in (0, L) \times (0, \infty)$$

with boundary conditions $u(0, t) = 0$, $u(L, t) = 0$, $t > 0$ and with initial conditions $u(x, 0) = 0$, $u_t(x, 0) = 0$.

Example 4.3: Find $u(x, t)$ such that $u_{tt} - u_{xx} = t \sin\left(\frac{\pi x}{L}\right)$, $(x, t) \in (0, L) \times (0, \infty)$
with boundary conditions $u(0, t) = 0$, $u(L, t) = 0$, $t > 0$ and with initial
conditions $u(x, 0) = 0$, $u_t(x, 0) = 0$, $x \in (0, L)$.

Solution: Suppose $U(x, t, s)$ is a solution to the user-defined problem:

$$U_{tt} - U_{xx} = 0, \quad (x, t) \in (0, L) \times (0, \infty).$$

with boundary conditions $U(0, t) = 0$, $U(L, t) = 0$, $t > 0$

and with initial conditions $U(x, 0) = 0$, $U_t(x, 0) = s \cdot \sin\left(\frac{\pi x}{L}\right)$, $x \in (0, L)$, $s > 0$.

The solution is given by

$$U(x, t, s) = \sum_{n=1}^{\infty} \left[A_n \cos\left(\frac{n\pi t}{L}\right) + B_n \sin\left(\frac{n\pi t}{L}\right) \right] \sin\left(\frac{n\pi x}{L}\right).$$

with
$$A_n = \frac{2}{L} \int_0^L F_1(x) \sin\left(\frac{n\pi x}{L}\right) dx$$
$$= 0, \quad n = 1, 2, 3\ldots\ldots$$

$$B_n = \frac{2}{n\pi} \int_0^L F_2(x) \sin\left(\frac{n\pi x}{L}\right) dx, \quad n = 1, 2, 3\ldots\ldots$$

$$B_n = \frac{2}{n\pi} \int_0^L s \sin\left(\frac{\pi x}{L}\right) \sin\left(\frac{n\pi x}{L}\right) dx.$$

Thus, $B_1 = \frac{sL}{\pi}$ and $B_n = 0$, $n \neq 1$, hence

$$U(x, t, s) = \frac{sL}{\pi} \sin\left(\frac{\pi x}{L}\right) \sin\left(\frac{\pi t}{L}\right).$$

Then, the solution $u(x, t)$ is

$$u(x, t) = \int_0^t U(x, t - \theta, \theta) d\theta$$

$$u(x, t) = \int_0^t \frac{\theta L}{\pi} \sin\left(\frac{\pi x}{L}\right) \sin\left(\frac{\pi (t - \theta)}{L}\right) d\theta$$

$$= \frac{L}{\pi} \sin\left(\frac{\pi x}{L}\right) \int_0^t \theta \sin\left(\frac{\pi (t - \theta)}{L}\right) d\theta$$

$$= \frac{L}{\pi} \sin\left(\frac{\pi x}{L}\right) \left[\theta \left[- \cos\left(\frac{\pi (t - \theta)}{L}\right)\left(\frac{-L}{\pi}\right) \right] - \int_0^t - \cos\left(\frac{\pi (t - \theta)}{L}\right)\left(\frac{-L}{\pi}\right) d\theta \right]$$

$$= \frac{L}{\pi} \sin\left(\frac{\pi x}{L}\right) \left[\cos\left(\frac{\pi (t - \theta)}{L}\right)\left(\frac{\theta L}{\pi}\right) + \sin\left(\frac{\pi (t - \theta)}{L}\right)\left(\frac{L}{\pi}\right)^2 \right]_0^t$$

$$u(x, t) = \frac{L}{\pi} \sin\left(\frac{\pi x}{L}\right) \left[\left(\frac{Lt}{\pi}\right) - \left(\frac{Lt}{\pi}\right)\cos\left(\frac{\pi t}{L}\right) - \sin\left(\frac{\pi t}{L}\right)\left(\frac{L}{\pi}\right)^2 \right].$$

4.2 ONE-DIMENSIONAL HEAT EQUATION

We shall solve the one-dimensional heat equation $\frac{\partial u}{\partial t} = c^2 \frac{\partial^2 u}{\partial x^2}$ for some practical approach, initial conditions, and boundary condition. We assume that two ends $x = 0$ and $x = L$ of the rod are insulated. Therefore, the temperature at two ends $x = 0$ and $x = L$ of the rod is zero. So that, boundary conditions are $u(0, t) = 0$, $u(L, t) = 0$ for all t and the initial temperature in the rod is $f(x)$.

We shall determine the solution of the temperature $u(x, t)$ of the heat equation which satisfying initial and boundary conditions.

Let us assume, $u(x, t) = X(x)T(t)$ be a solution to the heat equation.

Hence, it satisfies the heat equation.

Differentiate $u(x, t) = X(x)T(t)$ with respect to x and t

$$\frac{\partial u}{\partial x} = X'(x)T(t), \quad \frac{\partial u}{\partial t} = X(x)T'(t)$$

$$\frac{\partial^2 u}{\partial t^2} = X(x)T''(t).$$

Substituting above derivatives in $\frac{\partial u}{\partial t} = c^2 \frac{\partial^2 u}{\partial x^2}$

$$X(x)T'(t) = c^2 X''(x)T(t)$$

separates the variables

$$\frac{T'(t)}{c^2 T(t)} = \frac{X''(x).}{X(x)}$$

Since x and t are independent variables; therefore, $\frac{T'(t)}{c^2 T(t)} = \frac{X''(x)}{X(x)}$ can hold only when each side equal to some constant, say k

$$\frac{T'(t)}{c^2 T(t)} = \frac{X''(x)}{X(x)} = k$$

$$\frac{X''(x)}{X(x)} = k \text{ and } \frac{T'(t)}{c^2 T(t)} = k$$

$$\therefore X''(x) - kX(x) = 0 \text{ and } T'(t) - kc^2 T(t) = 0$$

$$\therefore D^2 X - kX = 0 \text{ and } DT - kc^2 T = 0.$$

Case 1: When k is positive
Let $k = p^2$
Then, $\therefore D^2 X - p^2 X = 0$

$$\therefore D^2 - p^2 = 0$$

The auxiliary equation is

$$\therefore m^2 - p^2 = 0$$

$$\therefore m = \pm p$$

$$X(x) = c_1 e^{px} + c_2 e^{-px}$$

For, $D - p^2 c^2 = 0$

$$m = p^2 c^2 t$$

$$T(t) = c_3 e^{c^2 p^2 t}.$$

Then, the solution is

$$u(x, t) = (c_1 e^{px} + c_2 e^{-px}) c_3 e^{c^2 p^2 t}.$$

Case 2: when k is negative
Let $k = -p^2$
Then, $\therefore D^2 X + p^2 X = 0$

$$\therefore D^2 + p^2 = 0$$

The auxiliary equation is

$$\therefore m^2 + p^2 = 0$$

$$\therefore m = \pm pi$$

$$X(x) = c_1 \cos px + c_2 \sin px$$

For, $D + p^2 c^2 = 0$

$$m = -p^2 c^2 t$$

$$T(t) = c_3 e^{-c^2 p^2 t}$$

Then the solution is

$$u(x, t) = (c_1 \cos px + c_2 \sin px) c_3 e^{-c^2 p^2 t}$$

Case 3: When $k = 0$
Let $k = 0$
Then, $\therefore D^2 X = 0$

$$\therefore D^2 = 0$$

The auxiliary equation is

$$\therefore m^2 = 0$$

$$\therefore m = 0, 0$$

$$X(x) = c_1 + c_2 x$$

For, $D = 0$

$$m = 0$$

$$T(t) = c_3.$$

Then, the solution is $u(x, t) = (c_1 + c_2 x)c_3$.

We have to choose that solution which is consistent with the physical nature of the problem. Since we are dealing with a problem with heat conduction, u must be transient. i.e. u is too decrease with the increase of time t.

Accordingly, $u(x, t) = (c_1 \cos px + c_2 \sin px)c_3 e^{-c^2 p^2 t}$ is the only suitable solution to the heat equation.

From the boundary condition $u(0, t) = (c_1 \cos p0 + c_2 \sin p0)c_3 e^{-c^2 p^2 t} = 0$

$$(c_1)c_3 e^{-c^2 p^2 t} = 0$$

$$\therefore c_1 = 0, \ c_3 \neq 0, \ e^{-c^2 p^2 t} \neq 0.$$

Hence, the solution reduced into $u(x, t) = (c_2 \sin px)c_3 e^{-c^2 p^2 t}$

$$\therefore u(L, t) = (c_2 \sin pL)c_3 e^{-c^2 p^2 t} = 0$$

$$\therefore c_2 \neq 0, \ c_3 \neq 0, \ e^{-c^2 p^2 t} \neq 0, \ \sin pL = 0$$

$$\therefore pL = n\pi, \text{ where, } n \in z$$
$$\therefore p = \frac{n\pi}{L}.$$

Therefore, the solution reduced into

$$u(x, t) = \left(c_2 \sin\left(\frac{n\pi x}{L}\right)\right)c_3 e^{-\frac{n^2 c^2 \pi^2}{L^2} t}$$

so by superposition theorem

$$u(x, t) = \sum_{n=1}^{\infty} c_n \left(\sin\left(\frac{n\pi x}{L}\right)\right) e^{-\frac{n^2 c^2 \pi^2}{L^2} t}.$$

For the initial condition,

$$u(x, 0) = \sum_{n=1}^{\infty} c_n \left(\sin\left(\frac{n\pi x}{L}\right) \right) = f(x)$$

$$f(x) = \sum_{n=1}^{\infty} c_n \left(\sin\left(\frac{n\pi x}{L}\right) \right)$$

which is a half-range sine series, where,

$$c_n = \frac{2}{L} \int_0^L f(x) \sin\left(\frac{n\pi x}{L}\right) dx, \quad n = 1, 2, 3...$$

Example 4.4: Determine the solution of the one-dimensional heat equation $\frac{\partial u}{\partial t} = c^2 \frac{\partial^2 u}{\partial x^2}$, where the boundary condition is $u(0, t) = u(L, t) = 0$, $t > 0$ and the initial condition is $u(x, 0) = x$, L being the length $0 < x < L$.

Solution: The solution of the heat equation $\frac{\partial u}{\partial t} = c^2 \frac{\partial^2 u}{\partial x^2}$ satisfying the given boundary condition is

$$u(x, t) = \sum_{n=1}^{\infty} c_n \left(\sin\left(\frac{n\pi x}{L}\right) \right) e^{-\frac{n^2 c^2 \pi^2}{L^2} t}.$$

Applying the boundary condition

When $t = 0 \Rightarrow u(x, 0) = x$

$$x = \sum_{n=1}^{\infty} c_n \left(\sin\left(\frac{n\pi x}{L}\right) \right).$$

This is the half-range Fourier series for the function $f(x) = x$ in the range $(0, L)$

where $c_n = \frac{2}{L} \int_0^L x \sin\left(\frac{n\pi x}{L}\right) dx$

$$c_n = \frac{2}{L} \left[x \left(\frac{-\cos\left(\frac{n\pi x}{L}\right)}{\frac{n\pi}{L}} \right) - \left(\frac{-\sin\left(\frac{n\pi x}{L}\right)}{\left(\frac{n\pi}{L}\right)^2} \right) \right]_0^L$$

$$= \frac{2}{L} \left[L \left(-\cos(n\pi) \frac{L}{n\pi} \right) \right]$$

$$c_n = -\frac{2L}{n\pi}(-1)^n$$

$$f(x) = \sum_{n=1}^{\infty}\left(-\frac{2L}{n\pi}(-1)^n\right)\sin\left(\frac{n\pi x}{L}\right).$$

Example 4.5: A homogeneous bar of conducting material of length 100 cm has its ends kept at zero temperature and the temperature initially is

$$u(x,0) = \begin{cases} x, & 0 \le x \le 50 \\ 100 - x, & 50 \le x \le 100 \end{cases}.$$

Find the temperature $u(x, t)$ at any time.

Solution: Let the equation for the conduction of heat be $\frac{\partial u}{\partial t} = c^2\frac{\partial^2 u}{\partial x^2}$.

Then the solution is

$$u(x,t) = \sum_{n=1}^{\infty} c_n\left(\sin\left(\frac{n\pi x}{L}\right)\right)e^{-\frac{n^2 c^2 \pi^2}{L^2}t}$$

$$\therefore u(x,0) = \sum_{n=1}^{\infty} c_n\left(\sin\left(\frac{n\pi x}{L}\right)\right)$$

where $c_n = \frac{2}{L}\int_0^L f(x)\sin\left(\frac{n\pi x}{L}\right)dx$ and $L = 100$

$$c_n = \frac{2}{100}\int_0^{100} f(x)\sin\left(\frac{n\pi x}{100}\right)dx$$

$$= \frac{2}{100}\left[\int_0^{50} x\sin\left(\frac{n\pi x}{100}\right)dx + \int_{50}^{100} (100-x)\sin\left(\frac{n\pi x}{100}\right)dx\right]$$

$$= \frac{2}{100}\left[x\left(-\cos\left(\frac{n\pi x}{100}\right)\right)\left(\frac{100}{n\pi}\right) - \left(-\sin\left(\frac{n\pi x}{100}\right)\left(\frac{100}{n\pi}\right)^2\right)\right]_0^{50}$$

$$+ \left[(100-x)\left(-\cos\left(\frac{n\pi x}{100}\right)\right)\left(\frac{100}{n\pi}\right) - (-1)\left(-\sin\left(\frac{n\pi x}{100}\right)\left(\frac{100}{n\pi}\right)^2\right)\right]_{50}^{100}$$

$$= \frac{2}{100}\left[50\left(-\cos\left(\frac{n\pi}{2}\right)\right)\left(\frac{100}{n\pi}\right) - \left(-\sin\left(\frac{n\pi}{2}\right)\left(\frac{100}{n\pi}\right)^2\right) + \left(\frac{5000}{n\pi}\right)\right]$$

$$+ \left[-50\left(-\cos\left(\frac{n\pi}{2}\right)\right)\left(\frac{100}{n\pi}\right) + \left(\sin\left(\frac{n\pi}{2}\right)\left(\frac{100}{n\pi}\right)^2\right)\right]$$

$$= \frac{2000}{100n\pi}\left[2\left(\sin\left(\frac{n\pi}{2}\right)\left(\frac{10}{n\pi}\right)\right) + 5\right].$$

Example 4.6: A rod of 30-cm long has its ends. A and B are kept at 20°C and 80°C, respectively, until steady-state condition prevails. The temperature at each end is suddenly reduced to 0°C and kept so. Find the resulting temperature $u(x, t)$ from end A.

Solution: Here, the temperature satisfies the heat equation,

$$\frac{\partial u}{\partial t} = c^2 \frac{\partial^2 u}{\partial x^2}.$$

Now, the sudden change of the temperature at the end B depends only upon x and not on t.

Thus, the heat equation becomes

$$\frac{\partial^2 u}{\partial x^2} = 0$$

$$\Rightarrow \frac{\partial u}{\partial t} = a$$
$$\Rightarrow u = ax + b$$

Since $u = 20$ for $x = 0$, we get

$$\therefore 20 = a(0) + b$$

$$\therefore b = 20.$$

So, $\therefore u = ax + 20$.

Also since $u = 80$ and $x = 60$ we get

$$\therefore 80 = a(30) + 20$$

$$\therefore 60 = a(30)$$

$$\therefore a = 2.$$

So, $u(x) = 2x + 20$.

Thus, the initial temperature is $u(x, 0) = 2(x + 10)$.

The solution of the heat equation is given by

$$u(x, t) = \sum_{n=1}^{\infty} c_n \sin\left(\frac{n\pi x}{L}\right) e^{-\frac{n^2\pi^2 c^2}{L^2}}$$

where $c_n = \frac{2}{L} \int_0^L f(x) \sin\left(\frac{n\pi x}{L}\right) dx$

$$= \frac{4}{30} \int_0^{30} (x + 10) \sin\left(\frac{n\pi x}{30}\right) dx$$

$$= \frac{4}{30}\left[(x + 10)\left(-\cos\left(\frac{n\pi x}{30}\right)\right)\left(\frac{30}{n\pi}\right) - \left(-\sin\left(\frac{n\pi x}{30}\right)\left(\frac{30}{n\pi}\right)^2\right)\right]_0^{30}$$

$$= \frac{4}{30}\left[40(-\cos(n\pi))\left(\frac{30}{n\pi}\right) - 10(-\cos(0))\left(\frac{30}{n\pi}\right)\right]$$

$$= \frac{40}{n\pi}[4(-1)^{n+1} + 1].$$

Hence, the solution of the heat equation is

$$u(x, t) = \sum_{n=1}^{\infty} \frac{40}{n\pi}[4(-1)^{n+1} + 1]\sin\left(\frac{n\pi x}{30}\right) e^{-\frac{n^2\pi^2 c^2}{900}}.$$

4.3 LAPLACE'S EQUATION

One of the most important partial differential equations in physics and engineering applications is **Laplace's equation**, given by $\nabla^2 u = u_{xx} + u_{yy} + u_{zz} = 0$.

Here, x, y, z are Cartesian coordinates in space. The expression $\nabla^2 u$ is called the **Laplacian** of u. The theory of the solutions of the Laplace equation is called a **potential theory**. Solutions that have continuous second partial derivatives are known as **harmonic functions**.

Laplace's equation occurs mainly in gravitation, electrostatics, steady-state heat flow, and fluid flow.

Practical problems involving Laplace's equation are boundary value problems in a region T in space with boundary surface S. Such problems can be grouped into three types:

I.   **First boundary value problem or Dirichlet problem** if u is prescribed on S.
II.   **Second boundary value problem** or **Neumann problem** if the normal a derivative is prescribed on S.
III. **Third** or **mixed boundary value problem** or **Robin problem** if u is prescribed on a portion of S and the remaining portion of S.

In general, when we want to solve a boundary value problem, we have to first select the appropriate coordinates in which the boundary surface S has a simple representation.

4.3.1 LAPLACIAN IN CYLINDRICAL COORDINATES

The first step in solving a boundary value problem is generally the introduction of coordinates in which the boundary surface S has a simple representation. Cylindrical symmetry (a cylinder as a region T) calls for cylindrical coordinates r, θ, z related to x, y, z by $x = r \cos \theta$, $y = r \sin \theta$, $z = z$.

So, from the Laplace equation, we can write,

$$\nabla^2 u = u_{rr} + \frac{1}{r} u_r + \frac{1}{r^2} u_{\theta\theta} + u_{zz}$$

4.3.2 LAPLACIAN IN SPHERICAL COORDINATES

Spherical symmetry (a ball as region T bounded by a sphere S) requires spherical coordinates r, θ, ϕ related to x, y, z by $x = r \cos \theta \sin \phi$, $y = r \sin \theta \sin \phi$, $z = r \cos \phi$.

Using the concept of chain rule, we obtain spherical coordinates

$$\nabla^2 u = u_{rr} + \frac{2}{r} u_r + \frac{1}{r^2} u_{\phi\phi} + \frac{\cot\phi}{r^2} u_\phi + \frac{1}{r^2 \sin^2 \phi} u_{\theta\theta}.$$

Sometimes, it can be written as

$$\therefore \nabla^2 u = \frac{1}{r^2} \left[\frac{\partial}{\partial r} r^2 \left(\frac{\partial u}{\partial r} \right) + \frac{1}{\sin\phi} \frac{\partial}{\partial \phi} \left(\sin\phi \frac{\partial u}{\partial \phi} \right) + \frac{1}{\sin^2 \phi} \frac{\partial^2 u}{\partial \phi^2} \right].$$

Example 4.7: Solve the following Dirichlet problem in spherical coordinates:

$$\therefore \nabla^2 u = \frac{1}{r^2}\left[\frac{\partial}{\partial r}r^2\left(\frac{\partial u}{\partial r}\right) + \frac{1}{\sin\phi}\frac{\partial}{\partial\phi}\left(\sin\phi\frac{\partial u}{\partial\phi}\right) + \frac{1}{\sin^2\phi}\frac{\partial^2 u}{\partial\phi^2}\right]$$

with the condition

$$u(R,\ \phi) = f(\phi),$$

$$\min_{r\to\infty} u(R,\ \phi) = 0.$$

Solution: Substitute $u(r,\ \phi) = F(r)G(\phi)$ into

$$\nabla^2 u = \frac{1}{r^2}\left[\frac{\partial}{\partial r}r^2\left(\frac{\partial u}{\partial r}\right) + \frac{1}{\sin\phi}\frac{\partial}{\partial\phi}\left(\sin\phi\frac{\partial u}{\partial\phi}\right) + \frac{1}{\sin^2\phi}\frac{\partial^2 u}{\partial\phi^2}\right].$$

Multiply r^2, making the substitution and then dividing by FG, we obtain

$$\frac{1}{F}\frac{d}{dr}\left(r^2\frac{dF}{dr}\right) = \frac{-1}{G\sin\phi}\frac{d}{d\phi}\left(\sin\phi\frac{dG}{d\phi}\right).$$

By the usual argument, both sides must be equal to a constant k. Thus, we get the two

ODEs

$$\frac{1}{F}\frac{d}{dr}\left(r^2\frac{dF}{dr}\right) = k \text{ or } r^2\frac{d^2F}{dr^2} + 2r\frac{dF}{dr} = kF$$

$$\frac{1}{\sin\phi}\frac{d}{d\phi}\left(\sin\phi\frac{dG}{d\phi}\right) + kG = 0.$$

Take $k = n(n+1)$ in $r^2\frac{d^2F}{dr^2} + 2r\frac{dF}{dr} = kF$, we obtain

$$r^2F'' + 2rF' - n(n+1)F = 0.$$

The above equation is the **Euler–Cauchy Equation**.

Hence, the roots are $m = n$, $m = -n - 1$.

Therefore, the solution is $F_n(r) = r^n$ and $F_n^*(r) = \frac{1}{r^{n+1}}$.

Now, we solve $\frac{1}{\sin\phi}\frac{d}{d\phi}\left(\sin\phi\frac{dG}{d\phi}\right) + kG = 0$, $\cos\phi = w$, we have $\sin^2\phi = 1 - w^2$.

And $\frac{d}{d\phi} = \frac{d}{dw}\frac{dw}{d\phi} = -\sin\phi\frac{d}{dw}$.

Take $k = n(n + 1)$ in $\frac{1}{\sin\phi}\frac{d}{d\phi}\left(\sin\phi\frac{dG}{d\phi}\right) + kG = 0$,

$$\frac{d}{dw}\left[(1 - w^2)\frac{dG}{dw}\right] + n(n + 1)G = 0.$$

Which can be written as

$$(1 - w^2)\frac{d^2G}{dw^2} - 2w\frac{dG}{dw} + n(n + 1)G = 0.$$

The above equation is called **Legendre's equation**.

For integer $n = 0, 1, 2....$ the **Legendre polynomials** are

$$G = P_n(w) = P_n(\cos\phi), \quad n = 0, 1, ...$$

The above equation is the solution of **Legendre's equation**.

We thus obtain the following two sequences of the solution $u = FG$ of the Laplace equation

$$\nabla^2 u = \frac{1}{r^2}\left[\frac{\partial}{\partial r}r^2\left(\frac{\partial u}{\partial r}\right) + \frac{1}{\sin\phi}\frac{\partial}{\partial\phi}\left(\sin\phi\frac{\partial u}{\partial\phi}\right) + \frac{1}{\sin^2\phi}\frac{\partial^2 u}{\partial\phi^2}\right],$$

with constant A_n and B_n, $n = 0, 1, 2...$

$$(a)u_n(r, \phi) = A_n r^n P_n\cos\phi, \quad (b)u_n^*(r, \phi) = \frac{B_n}{r^{n+1}}P_n\cos\phi.$$

EXERCISES

Q1 A tightly stretched string fastened between two pints at a distance 10 cm apart. If initially, the string has a shape $u = 4\sin^3\left(\frac{\pi x}{10}\right)$, and then it is released to rest from this position. Determine the vertical displacement of any point x from one end at any time $\left(\text{given } c^2 = \frac{1}{9}\right)$.

Q2 A string of length fixed l at both ends an initially at rest and initial $u = k(lx - x^2)$; $0 \le x \le l$. Find the displacement of the string at any time t.

Q3 Let $u(x, t)$ be the required temperature of the rod which satisfies one-dimensional heat flow equation $\frac{\partial u}{\partial t} = c^2 \frac{\partial^2 u}{\partial x^2}$

Subject to following conditions: $u(0, t) = u(l, t) = 0$, $t > 0$
$u(x, 0) = f(x)$; $0 < x < l$

Q4 A rod of length l with insulated sides is initially at a uniform temperature u_0. Its ends are suddenly cooled to $0°C$ and are kept at that temperature. Find the temperature $u(x, t)$.

Q5 Find the temperature in an insulated bare of length L whose ends are kept at temperature 0 assuming that the initial temperature is

$$f(x) = \begin{cases} x & \text{if } 0 < x < \frac{L}{2} \\ L - x & \text{if } \frac{L}{2} < x < L \end{cases}$$

Q6 Solve $\frac{\partial^2 u}{\partial x^2} + \frac{\partial^2 u}{\partial y^2} = 0$ with conditions given that (i) $u(0, y) = 0$ for all values of y;

(ii) $u(\pi, y) = 0$ for all values of x (iii) $u(x, \infty) = 0$ in $0 < x < \pi$ (iv) $u(x, 0) = u_0$
in $0 < x < \pi$

Q7 Find the solution of wave equation $\frac{\partial^2 u}{\partial t^2} = c^2 \frac{\partial^2 u}{\partial x^2}$ under the conditions

(i) $u(0, t) = 0$, for all t
(ii) $u(1, t) = 0$, for all t

(iii) $u(x, 0) = f(x) = \begin{cases} 2kx & \text{if } 0 < x < \frac{1}{2} \\ 2k(1 - x) & \text{if } \frac{1}{2} < x < 1 \end{cases}$

(iv) $\left(\frac{\partial u}{\partial t}\right)_{t=0} = g(x) = 0$

Q8 To solve the heat equation $\frac{\partial u}{\partial t} = c^2 \frac{\partial^2 u}{\partial x^2}$. How many initial and boundary conditions require?

Q9 Solve $\frac{\partial^2 u}{\partial x^2} + \frac{\partial^2 u}{\partial y^2} = 0$ for $0 < x < \pi, 0 < y < \pi$ with conditions given:

$u(0, y) = u(\pi, y) = u, u(x, 0) = \sin^2 x$

Q10 Solve the equation $\frac{\partial u}{\partial t} = k \frac{\partial^2 u}{\partial x^2}$ for the condition of heat along a rod without radiation subject to the condition

(i) $\frac{\partial u}{\partial t} = 0$ for $x = 0$ and $x = l$
(ii) $u = lx - x^2$ at $t = 0$ and for all x

ANSWERS

1 $u(x, t) = 3\sin\left(\frac{\pi x}{10}\right)\cos\left(\frac{\pi t}{30}\right) - \sin\left(\frac{3\pi x}{10}\right)\cos\left(\frac{\pi t}{10}\right)$

2 $u = \frac{8kl^2}{\pi^3}\sum_{n=1}^{\infty}\frac{1}{(2n-1)^3}\sin\left[\frac{(2n-1)\pi x}{l}\right]\cos\left[\frac{(2n-1)\pi x}{l}\right]$

3 $u(x, t) = \sum_{n=1}^{\infty}A_n\sin\left(\frac{n\pi x}{L}\right)e^{-\frac{c^2 n^2 \pi^2}{l^2}t}$, where $A_n = \frac{2}{l}\int_0^l f(x)\sin\left(\frac{n\pi x}{L}\right)dx$

4 $u(x, t) = \frac{4\mu_0}{\pi}\sum_{m=1}^{\infty}\frac{1}{(2m-1)}\sin\left[\frac{(2m-1)\pi x}{l}\right]e^{\frac{c^2(2m-1)^2\pi^2}{l^2}t}$

5 $u(x, t) = \frac{4l}{\pi^2}\left[\sin\left(\frac{\pi x}{l}\right)e^{\frac{c^2\pi^2}{l^2}t} - \frac{1}{9}\sin\left(\frac{3\pi x}{l}\right)e^{-\frac{9c^2\pi^2}{l^2}t}\right]$

6 $u(x, t) = \frac{4u_0}{\pi}\left[e^{-y}\sin(x) + \frac{1}{3}e^{-3y}\sin(3x) + \frac{1}{5}e^{-5y}\sin(5x) + \dots\dots\right]$

7 $u(x, t) = \frac{8k}{\pi^2}\begin{bmatrix}\frac{1}{l^2}\sin(\pi x)\cos(\pi ct) - \frac{1}{3^2}\sin(3\pi x)\cos(3\pi ct) \\ + \frac{1}{5^2}\sin(5\pi x)\cos(5\pi ct) - \dots\dots\end{bmatrix}$

8 One initial condition and a two boundary condition.

9 $u(x, y) = -\frac{8}{\pi}\sum_{n=1,3,5}^{\infty}\frac{\sin nx\,\sinh n(x-y)}{n(n^2-4)\sinh n\pi}$

10 $u(x, t) = \frac{l^2}{6} - \frac{l^2}{\pi^2}\sum_{m=1}^{\infty}\frac{1}{m^2}\cos\left[\frac{2m\pi x}{l}\right]e^{\frac{4m^2 k\pi^2}{l^2}t}$

MULTIPLE-CHOICE QUESTIONS

1 In a general second-order linear partial differential equation with two independent variables

$$A\frac{\partial^2 u}{\partial x^2} + B\frac{\partial^2 u}{\partial x \partial y} + C\frac{\partial^2 u}{\partial y^2} + D = 0$$

where A, B, C are functions of x and y, and D is a function of x, y, $\frac{\partial u}{\partial x}$, $\frac{\partial u}{\partial y}$, then the partial differential equation is parabolic if
(A) $B^2 - 4AC < 0$ (B) $B^2 - 4AC > 0$
(C) $B^2 - 4AC = 0$ (D) $B^2 - 4AC \neq 0$
Answer: (C) $B^2 - 4AC = 0$.

2 A partial differential equation requires
(A) exactly one independent variable   (B) two or more independent variables
(C) more than one dependent variable (D) equal number of dependent and
                                                                  independent variables
Answer: (B) two or more independent variables.

3 Using substitution, which of the following equations are solutions to the partial differential equation?

$$\frac{\partial^2 u}{\partial x^2} = 9\frac{\partial^2 u}{\partial y^2}$$

(A) $\cos(3x - y)$ (B) $x^2 + y^2$
(C) $\sin(3x - 3y)$ (D) $e^{-3\pi x}\sin(\pi y)$
Answer: (A) $\cos(3x - y)$.

4 The following is true for the following partial differential equation used in non-linear mechanics known as the Korteweg–de Vries equation.

$$\frac{\partial w}{\partial t} + \frac{\partial^3 w}{\partial x^3} - 6w\frac{\partial w}{\partial x} = 0$$

(A) linear; third order   (B) non-linear; third order
(C) linear; first order   (D) non-linear; first order
Answer: (B) non-linear; third order.

5 The region in which the following equation $x^3\frac{\partial^2 u}{\partial x^2} + 27\frac{\partial^2 u}{\partial y^2} + 3\frac{\partial^2 u}{\partial x \partial y} + 5u = 0$ acts as an elliptic equation is
(A) $x > (\frac{1}{12})^{1/3}$ (B) $x < (\frac{1}{12})^{1/3}$
(C) for all values of x (D) $x = (\frac{1}{12})^{1/3}$
Answer: (A) $x > (\frac{1}{12})^{1/3}$.

6 The finite difference approximation of $\frac{\partial^2 u}{\partial x^2}$ in the elliptic equation

$\frac{\partial^2 u}{\partial x^2} + \frac{\partial^2 u}{\partial y^2} = 0$ at (x, y) can be approximated as

(A) $\frac{\partial^2 u}{\partial x^2} \cong \frac{u(x+\Delta x, y) - 2u(x, y) + u(x - \Delta x, y)}{(\Delta x)^2}$

(B) $\frac{\partial^2 u}{\partial x^2} \cong \frac{u(x+\Delta x, y) - u(x, y) + u(x - \Delta x, y)}{(\Delta x)^2}$

(C) $\frac{\partial^2 u}{\partial x^2} \cong \frac{u(x, y+\Delta y) - 2u(x, y) + u(x, y - \Delta y)}{(\Delta x)^2}$

(D) $\frac{\partial^2 u}{\partial x^2} \cong \frac{u(x+\Delta x, y) - u(x - \Delta x, y)}{2\Delta x}$

Answer: (A) $\frac{\partial^2 u}{\partial x^2} \cong \frac{u(x+\Delta x, y) - 2u(x, y) + u(x - \Delta x, y)}{(\Delta x)^2}$.

7 A partial differential equation has
   (A) one independent variable
   (B) two or more independent variables
   (C) more than one dependent variable
   (D) an equal number of dependent and independent variables
   Answer: (B) two or more independent variables.

8 Lagrange's linear equation is of the form
   (A) $Pp + Qq = R$ (B) $pp + qq = R$
   (C) $P + Qq = R$ (D) $Pp + Q = R$
   Answer: (A) $Pp + Qq = R$.

9 Eliminating the arbitrary constants a and b from $(x - a)^2 + (y - b)^2 + z^2 = c^2$, the partial differential equation formed is
   (A) $z^2(p^2 + q^2 + 1) = c^2$ (B) $z^2(p + q + 1) = c^2$
   (C) $z(p + q + 1) = c^2$ (D) None of these
   Answer: (A) $z^2(p^2 + q^2 + 1) = c^2$.

10 Eliminating the arbitrary constants a and b from $x^2 + y^2 + (z - c)^2 = a^2$, the partial differential equation formed is
   (A) $xp = yq$ (B) $xq = yp$
   (C) $z = pq$ (D) none of these
   Answer: (A) $xp = yq$.

11 The general solution of the equation $xp + yq = z$ is
   (A) $F\left(\frac{x}{y}, \frac{y}{z}\right) = 0$ (B) $F(xy, x + y) = 0$
   (C) $F\left(\frac{y}{x}, \frac{z}{y}\right) = 0$ (D) none of these
   Answer: (A) $F\left(\frac{x}{y}, \frac{y}{z}\right) = 0$.

12 The general solution of the equation $z = px + qy + p^2q^2$ is
   (A) $z = ax + by + c$ (B) $z = ax + by + x^2 + y^2$
   (C) $z = ax + by - a^2b^2$ (D) $z = ax + by + a^2b^2$
   Answer: (D) $z = ax + by + a^2b^2$.

FILL IN THE BLANKS

1 The partial differential equation $5\frac{\partial^2 z}{\partial x^2} + 6\frac{\partial^2 z}{\partial y^2} = xy$ is classified as _____.
Answer: Elliptic.

2 The partial differential equation $xy\frac{\partial z}{\partial x} = 5\frac{\partial^2 z}{\partial y^2}$ is classified as _____.
Answer: Parabolic.

3 For solving one-dimensional wave equation $\frac{\partial^2 u}{\partial t^2} = c^2 \frac{\partial^2 u}{\partial x^2}$, the number of initial conditions required is _____ and boundary conditions required is _____.
Answer: 2, 2.

4 The order and degree of the partial differential equation $x(\frac{\partial^3 z}{\partial x^3})^4 + y(\frac{\partial z}{\partial y})^5 = z$ is _____.
Answer: order = 3, degree = 4.

5 For partial differential equation, if $b^2 - 4ac = 0$ then the equation is called _____.
Answer: Parabolic.

Bibliography

Kachot, K. R. (2015). 'Advanced engineering mathematics', 7th Edition, 524–615, Mahajan Publication House.

Kreyszig, E. (2011). 'Advanced engineering mathematics' 10th Edition, 540–604, WILEY.

Ram, B. (2011). 'Engineering mathematics III', 8.1–8.31, Pearson Publication.

Sankara Rao, K. (2011). 'Introduction to partial differential equation', 3rd Edition, PHI Learning Private Limited.

Index

For Product Safety Concerns and Information please contact our EU
representative GPSR@taylorandfrancis.com
Taylor & Francis Verlag GmbH, Kaufingerstraße 24, 80331 München, Germany

9 780367 613235